STEM CELL BIOLOGY
Basic Concepts to Frontiers

S T U D E N T S   E D I T I O N

EDITED BY
Paolo Di Nardo and **Dinender K. Singla**

ISBN: 1466291079
ISBN-13: 9781466291072
Library of Congress Control Number: 2011916364

CreateSpace, North Charleston, SC

Acknowledgement

The editors would like to thank the contributing authors for all the work they have done in creating this book.

Contributing Authors

Michael Bauer
Cardiac Muscle Research Laboratory
Cardiovascular Division
Department of Medicine
Brigham and Women's Hospital
Harvard Medical School, Boston, MA

Ahmad F. Bayomy
Cardiac Muscle Research Laboratory
Cardiovascular Division
Department of Medicine
Brigham and Women's Hospital
Harvard Medical School, Boston, MA
University of Washington School of Medicine, Seattle, WA

Sergey V. Bushnev
Neuroscience & Orthopaedic Clinical Research Institute
Florida Hospital
Orlando, FL 32804

Luciana Carosella
Canadian-Italian Tissue Engineering Laboratory (CITEL)
St. Boniface Research Center, Winnipeg, Manitoba, Canada
BioLink Insitute, Link Campas Unversity, Rome, Italy

Paolo Di Nardo
Laboratorio di Cadiologia Molecolare e Cellulare
Dipartimento di Medicina Interna
Universita di Roma Tor Vergata, Roma, Italy
Japanese-Italian Tissue Engineering Laboratory (JITEL)
Tokyo Women's Medical University-Waseda University
Joint Institution for Advanced Biomedical Sciences (TWIns),
Tokyo, Japan
BioLink Insitute, Link Campas Unversity, Rome, Italy

Miguel Angel Esteban
Stem Cell and Cancer Biology Group
Key Laboratory of Regenerative Biology
South China Institute of Stem Cell Biology and Regenerative Medicine
Guangzhou Institutes of Biomedicine and Health
Chinese Academy of Sciences, Guangzhou 510530, China

Giancarlo Forte
Laboratorio di Cadiologia Molecolare e Cellulare
Dipartimento di Medicina Interna
Universita di Roma Tor Vergata, Roma, Italy
Biomaterials Center
International Center for Materials Nanoarchitectonics (MANA)
National Institute for Materials Science (NIMS)
Tsukuba, Japan

Carl Gregory
Institute for Regenerative Medicine at Scott and White Hospital
Texas A and M Health Science Center
Module C, 5701 Airport Road
Temple, TX, USA

Junjiu Huang
School of Life Sciences
Sun Yat-sen University
Guangzhou 510275, China
Department of Obstetrics and Gynecology
University of South Florida College of Medicine
Tampa, FL 33647, USA

Keiichi Katsumoto
Department of Stem Cell Biology
Institute of Molecular Embryology and Genetics (IMEG)
Kumamoto University
The Global COE Cell Fate Regulation Research and Education Unit
Kumamoto University
Honjo 2-2-1, Kumamoto 860-0811, Japan

David L. Keefe
Department of Obstetrics and Gynecology
University of South Florida College of Medicine
Tampa, FL 33647, USA

Shoen Kume
Department of Stem Cell Biology
Institute of Molecular Embryology and Genetics (IMEG)
Kumamoto University
The Global COE Cell Fate Regulation Research and Education Unit
Kumamoto University
Honjo 2-2-1, Kumamoto 860-0811, Japan

Ronglih Liao
Cardiac Muscle Research Laboratory
Cardiovascular Division
Department of Medicine
Brigham and Women's Hospital
Harvard Medical School, Boston, MA

Jinmao Liu
College of Life Sciences
Nankai University
Tianjin 300071, China

Lin Liu
Department of Obstetrics and Gynecology
University of South Florida College of Medicine
Tampa, FL 33647, USA
College of Life Sciences
Nankai University
Tianjin 300071, China

Almudena Martinez-Fernandez
Division of Cardiovascular Diseases
Department of Medicine
Mayo Clinic
Rochester, MN

Rika Miki
Department of Stem Cell Biology
Institute of Molecular Embryology and Genetics (IMEG)
Kumamoto University
The Global COE Cell Fate Regulation Research and Education Unit
Kumamoto University
Honjo 2-2-1, Kumamoto 860-0811, Japan

Marilena Minieri
Laboratorio di Cadiologia Molecolare e Cellulare
Dipartimento di Medicina Interna
Universita di Roma Tor Vergata, Roma, Italy
Japanese-Italian Tissue Engineering Laboratory (JITEL)
Tokyo Women's Medical University-Waseda University
Joint Institution for Advanced Biomedical Sciences (TWIns),
Tokyo, Japan
Canadian-Italian Tissue Engineering Laboratory (CITEL)
St. Boniface Research Center, Winnipeg, Manitoba, Canada

Timothy J. Nelson
Division of Cardiovascular Diseases
Department of Medicine
Mayo Clinic
Rochester, MN

Michel Puceat
INSERM
Evry France
INSERM University Paris Descartes UMR633
4, Rue Pierre Fontaine
91058 Evry, France

Crystal M. Rocher
Burnett School of Biomedical Sciences
College of Medicine
University of Central Florida
Orlando, FL 32817

Nobuaki Shiraki
Department of Stem Cell Biology
Institute of Molecular Embryology and Genetics (IMEG)
Kumamoto University
Honjo 2-2-1, Kumamoto 860-0811, Japan

Dinender K. Singla
Burnett School of Biomedical Sciences
College of Medicine
University of Central Florida
Orlando, FL 32817

Jeffrey L. Spees
Department of Medicine and Stem Cell Core
University of Vermont
Colchester, VT 05446

Sonia Stefanovic
INSERM
Evry France
INSERM University Paris Descartes UMR633
4, Rue Pierre Fontaine
91058 Evry, France

Kiminobu Sugaya
Burnet School of Biomedical Science
College of Medicine
University of Central Florida, Orlando, Florida, USA

Andre Terzic
Division of Cardiovascular Diseases
Department of Medicine
Mayo Clinic
Rochester, MN

Kahoko Umeda
Department of Stem Cell Biology
Institute of Molecular Embryology and Genetics (IMEG)
Kumamoto University
Honjo 2-2-1, Kumamoto 860-0811, Japan

Fang Wang
School of Life Sciences
Sun Yat-sen University
Guangzhou 510275, China
College of Life Sciences
Nankai University
Tianjin 300071, China

Zhong Wang
Cardiovascular Research Center
Massachusetts General Hospital
Harvard Medical School and Harvard Stem Cell Institute
Richard Simches Research Center
185 Cambridge Street, Boston, MA 02114

Satsuki Yamada
Division of Cardiovascular Diseases
Department of Medicine
Mayo Clinic
Rochester, MN

Xiaoying Ye
College of Life Sciences
Nankai University
Tianjin 300071, China

Ming Zhan
Department of Systems Medicine and Bioengineering
The Methodist Hospital Research Institute
Houston, TX, USA

Contents

Chapter 1

EMBRYONIC STEM CELL BIOLOGY

Sonia Stefanovic and Michel Puceat

For correspondence:
INSERM (National Institute for Health and medical Research),
Evry France
INSERM University Paris Descartes UMR633
4, Rue Pierre Fontaine
91058 Evry, France

E-mail: Michel.puceat@inserm.fr

1. What Is a Stem Cell?

The origin of the stem cell appellation emerged more than a century ago when a German scientist, Ernst Haeckel, a major supporter of Darwin's theory of evolution, made a number of phylogenetic trees, searching for common ancestors of organisms. He called these trees *Stammbäume* (family trees or stem trees). He then used the term *Stammzelle* (stem cell) to describe the unicellular ancestor organism from which he presumed all multicellular organisms derived [1]. The stem cell is a fascinating cell capable of self-renewal and differentiation in one (unipotent), several (multipotent), and all (pluripotent) embryonic, as well as extraembryonic (totipotent) cell lineages. *Self-renewal* indicates cell division that gives rise to at least one daughter cell (out of two) that is identical to the mother cell.

Stem cells can be divided into two broad categories: adult stem (somatic) cells and embryonic stem (ES) cells. To date, three pluripotent cell types have been established from mouse and human tissue: embryonic carcinoma (EC) cells, embryonic germ (EG) cells, and ES cells. EC cell lines were the first pluripotent stem cell lines to be established. They were derived from the undifferentiated stem cell component of germ cell tumors. The EC clones could be expanded continuously in culture but could also differentiate to produce derivatives of all three germ layers either in vitro or through teratocarcinoma formation. Compared with ES cells, however, these cells seem to have less differentiating capacity, and they are usually aneuploid and, therefore, not suitable for clinical applications. EG lines were derived from primordial germ cells in the genital ridges of the developing embryos, typically at five to nine weeks after fertilization in humans. They were also shown to be pluripotent.

2. What Is an Embryonic Stem Cell?

2.1 Embryonic origin

Cells in the early mammalian embryo have the potential to contribute to all tissue types in the body (defined as pluripotency). After fertilization, at the blastocyst stage, a hollow sphere of cells is formed that contains an outer cell layer and an inner cluster of cells termed the inner cell mass (ICM). Whereas the outer cells constitute the trophectoderm and subsequently give rise to the placenta and other supporting tissues (extraembryonic tissues), the ICM cells ultimately create all tissues in the body (embryonic tissues), as well as nontrophoblast structures (extraembryonic tissues) that support the embryo. They are, therefore, truly pluripotent. The term ES cells originated from the isolation in 1981 of pluripotent stem cell cultures from mouse blastocysts by Evans M. J. and Kaufman M. H. and independently by Martin G. R. (**Nobel Prize 2007**) [2, 3]. Embryonic or pluripotent stem cells proliferate forever in culture, never reaching the senescent crisis, the point at which a primary cell stops dividing after a finite number of divisions (the Hayflick limit) when telomeres reach critical lengths. Indeed, embryonic stem cells feature a high telomerase activity preventing this phenomenon. Stem cells provide a powerful model to study embryonic development, cellular differentiation, and organ maintenance. Given their potential of differentiation, the stem cells are also a potential source for organ regeneration.

Because of their origin in the early embryo, ES cells differ from other stem cells in their ability to retain the potential to generate derivatives of all three germ layers (endoderm, mesoderm, and ectoderm). ES cell lines were first established in mice. Later, pluripotent lines of monkey and human ES cells were also established [4–6]. Since their differentiation potency is similar to that of the murine ES cells, these cells hold hopes in regenerative medicine to resolve the problems of donor shortage limiting organ transplantation. They also represent a good model for the understanding of human development. Many efforts have been made to establish ES cell lines from other mammalian species, including rat (2006), pig (2007), mink (1993), cow (2005), horse (2002), sheep (1991), rabbit (1993), and dog (2006).

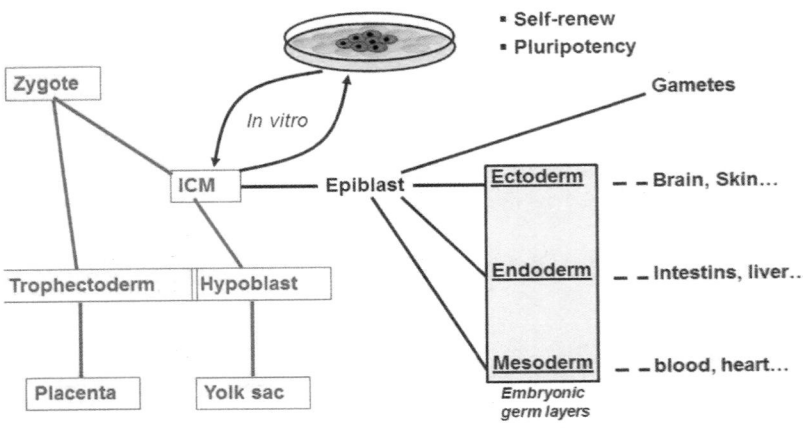

2.2 Protocol of derivation

Human ES cell lines are derived from the ICM cells of spare human blastocysts produced by in vitro fertilization for clinical purposes and donated by individuals after informed consent. The human ES cell lines are created in a manner similar to that of mouse and monkey ES cells. In this process, the outer trophectoderm layer of the blastocyst is selectively removed using anti–Human IgG antibodies added with the complement   (the complement improves the ability of antibodies to clear pathogens from an organism) (immunosurgery). The ICM cells are isolated and plated on a mitotically inactivated mouse embryonic fibroblast (MEF) feeder layer. MEF cells provide the ICM cells with a stromal surface onto which they can attach. The feeder cells release nutrients into the culture medium. To maintain an undifferentiated state, ES cells require a specific environment provided by the MEFs, but they also require the presence of morph genes such as *LIF* (*leukemia inhibitory factor*) and *FGF2* (*fibroblast growth factor 2*) for mouse and human ES cells, respectively. Cells form colonies that are mechanically isolated and replated until homogenous colonies appear. These colonies are selected, passaged, and expanded for the creation of ES cell lines.

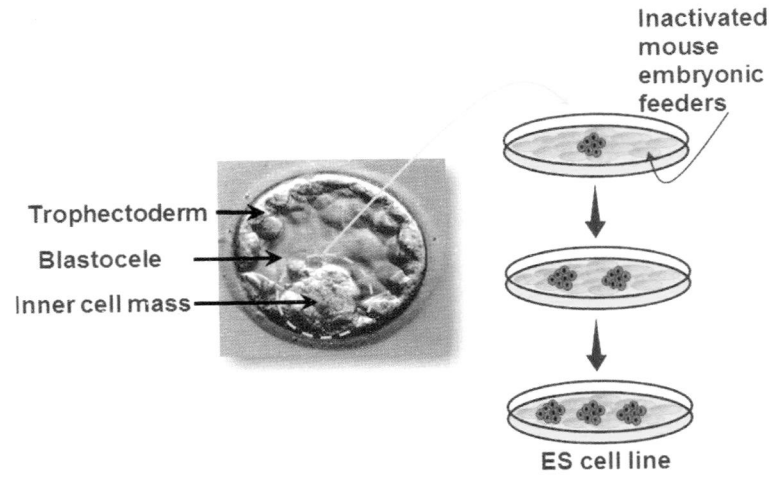

Trophectoderm

Blastocele

Inner cell mass

Inactivated mouse embryonic feeders

ES cell line

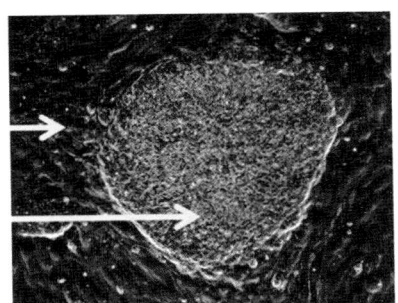

Feeders

Human ES cells

Human ES cells colonie

2.3 ES cell characteristics

ES cells are defined by the presence of several transcription factors and cell surface proteins associated with an undifferentiated state. The transcription factors *OCT4*, *NANOG*, and *SOX2* form the core regulatory network, which ensures the suppression of genes that lead to differentiation and the maintenance of pluripotency.

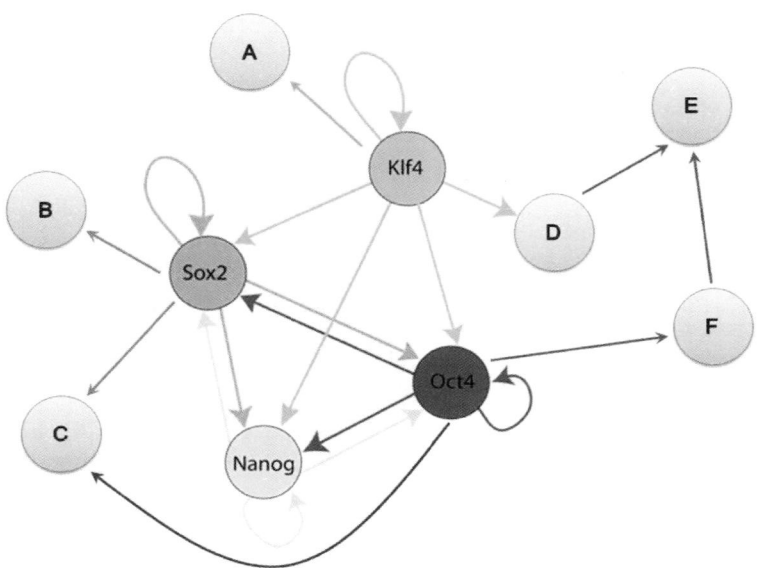

Genes that are required later in development are also repressed by epigenetic modifications. Recent studies have shown that epigenetic status is also an important criterion for ES cells. This epigenetic silencing is flexible during the early mammalian development.

The cell surface antigens most commonly used to identify hES cells are the glycolipids SSEA-3 and SSEA-4 and the keratan sulfate antigens Tra-1-60 and Tra-1-81. The molecular definition of a stem cell includes many more proteins and continues to be a topic of research.

ES cells can maintain their undifferentiated state and pluripotency in repeated subcultures. They are evaluated by their capacity to differentiate, which can be established traditionally using three different approaches:

1. Mouse ES cells can be retransferred into early mouse embryos, where they eventually give rise to all somatic cells of the chimeric embryo, including the germ

cells. Such a test cannot be applied to human ES cells for obvious ethical reasons.

2. The second approach relates to the demonstration that ES cells can differentiate to generate derivatives of all three germ layers in vivo. When ES cells are injected into immunodeficient mice, they form benign tumors called teratomas, containing advanced differentiated tissue types representing all three germ layers.

3. The third approach establishes ES cells pluripotency during in vitro differentiation. Both mouse and human ES cells, when removed from the MEF feeder layer and allowed to differentiate, can form three-dimensional cell aggregates, termed embryoid bodies (EBs), which contain tissue derivatives of endodermal, ectodermal, and mesodermal origin.

Normal karyotype is also an important characteristic of these cells. Indeed, ES cell lines have been shown to keep a stable diploid karyotype and to continuously express a high level of telomerase activity during long-term propagation in culture.

3. Differentiation

3.1 Definitions

The process of a cell differentiation can be divided into four steps: specification (exit from self-renewal), determination, patterning, and differentiation.

These events, which ensure the maintenance of cell pluripotency, are under the control of growth factors, or morphogens, including FGF, WnT, BMP, TGFβ, and Nodal, acting through intrasignaling pathways on gene transcription.

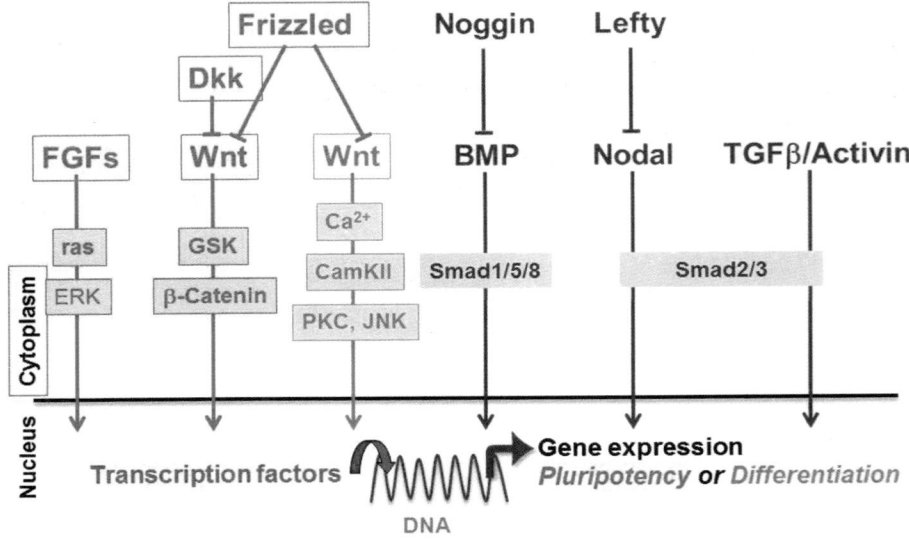

3.2 Potential of differentiation

Embryonic stem cells can give rise to the three primary germ layers (ectoderm, mesoderm, and endoderm) and, in turn, their derivatives. The limited availability of pre-implantation embryos in mice and, especially, humans has revealed the potential of ES cells in investigating the first steps of differentiation toward the three germ layers. The three-dimensional structure formed by the aggregation of ES cells, named embryoid bodies (EBs), was first set by Doetschman in 1985 [7]. Basically, grown in the absence of feeders and without cytokines (LIF or FGF2) to maintain their pluripotency in non-adherent surface or as hanging drops, ES cells are capable to spontaneously form spheroid aggregates recapitulating the first stages of differentiation. They first differentiate toward primitive endoderm (extraembryonic endoderm) and then form the three germ layers. This cell configuration turns out to be a very powerful aid in the study of differentiation toward any cell lineage in an embryonic-like environment. It should be noted that co-culture can in some cases also efficiently differentiate ES cells toward specific cell lineages. However, only EBs will allow ES cells to form most, if not all, embryonic cell lineages.

3.2.1 Differentiation toward ectoderm (neurons, skin)

For many years, neuroectodermal differentiation of ES cells has been considered a default differentiation. LIF-starved mouse and FGF2-starved human ES cells spontaneously express ecto-dermal genes and, more specifically, neuronal markers. How-ever, to guide their fate and improve spontaneous differentia-tion of the ectoderm within EBs, growth factors are mandatory. ES cells are allowed to form EBs. The EBs are differentiated to neuroepithelial cells in a serum-free culture medium and then plated on laminin-coated dishes. These cells express *nestin*, SOX2, *SOX1* (*SRY-related HMG-box*), and *Pax6* (*Paired box*).

Under this experimental condition, neural fate can be ob-served in clusters of elongated cells surrounding a central, small, and cell free area. These structures, named rosettes, express early neural markers such as *nestin* and *Musashi-1*. The neuroepithelial cells can be differentiated to neuronal and glial progenitors with forebrain, midbrain, hindbrain, and spinal cord prints using different growth factors [8]. The com-bination of Noggin, an inhibitor of BMP receptors, Dickkopf, an inhibitor of WnT pathway, and IGF [9] directs the fate of the progenitors toward retinal cells. Basic FGF (FGF2) and EGF stimulation gives rise to neural progenitors expressing *OLIG1* (*oligodendrocyte lineage gene*), A2B5 (*A2B5-reactive gan-glioside*), and *SOX10*, and further to oligodendrocytes ex-pressing O4, O1, MBP (*myelin basic protein*), PLP (*proteo lipid protein*), NGN2 (*NeuroGeNin 2*), HB9 (*HomeoBox gene 9*) and synapsin. The rosettes can be further differentiated into progenitors of motoneurons using retinoic acid and SHH (Sonic hedgehog) and then into motoneurons using BDNF, GDNF, and IGF. Mature motoneurons express *NKX6.1, OLIG2, NGN2, ISL1, CHAT (choline acetyl transferase), VACHT (vesicular acetylcholine transporter)*, and *FGF8 (fibroblast growth factor 8)*. SHH drives the rosettes towards dopaminergic neurons featuring expression of *MAP2 (microtubule associated protein 2), TH (tyroxine hydroxylase), AADC (aromatic L-amino acid decarboxylase), VMAT (vesicular monoamine transporter), NURR1 (nuclear receptor related-1)*, and *PTX3 (pentraxin-related 3)*. Several protocols have been designed to improve the differentiation protocols into specific neurons and these can be found in two recent reviews [8, 9].

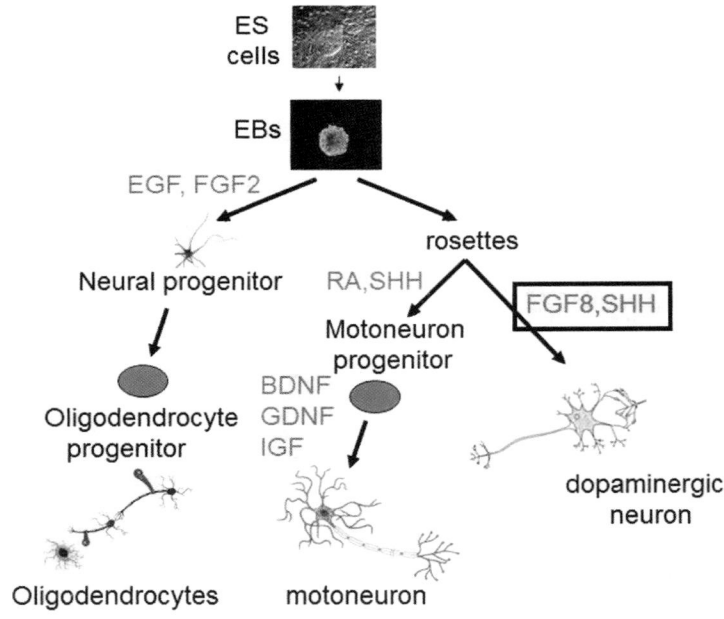

In 1996, F. Watt's group [10] reported that differentiation of ES cells within EBs could recapitulate embryonic epidermal differentiation. Expression of epidermal markers was observed. First, *K8* and *K18* genes, the first keratins to be expressed in simple epithelia during mouse development, were detected after fifteen days in EB culture. Then, keratinocytes expressing *K14* and later *K10* and involucrin were monitored in EBs. Thus, terminally differentiated keratinocytes can be obtained using the EBs culture.

More recently, a two-dimensional co-culture was used to derive epidermal cells from mouse ES cells. ES cells were plated on slides coated with extracellular matrix secreted by numerous primary cultures and cell lines of various origins. The most efficient keratinocyte induction was obtained when ES cells were seeded on matrix derived from human normal fibroblasts (HNF) and NIH-3T3 cells, both of mesenchymal origin. Compared to EB formation in which keratinocytes appear at day twenty-one, the co-culture protocol allows generation of *K14*-positive cells as early as eight days after induction. An improvement of epidermal differentiation was observed after addition of BMP4 in the culture [11]. *K14*-positive ectodermal cells have also been derived from human ES cells [12].

3.2.2 Differentiation toward mesoderm (blood cells, cardiac)

The first stage of differentiation of ES cells into the embryonic endoderm and mesoderm is the induction of the primitive streak marked by expression of *T* (*brachyury*), *GSC* (*Goosecoid*), and *MIXL1*.

In embryo, induction of the primitive streak is triggered by WnT and Activin/Nodal/TGFβ pathway [13]. The primitive streak segregates into the posterior and anterior region leading to the formation of the mesoderm and embryonic endoderm, respectively. Addition of BMPs to ES cells induces expression of the posterior primitive streak markers *HOXB1* and *MESP1* (*mesoderm posterior1*) while expression of the anterior

11

markers *CHRD* (*chordin*) and *CER1* (*Cerberus like 1*) remains low. The action of BMPs requires activation of both WnT and TGFβ pathways, as inhibitors of the latter (DKK and the TGFβ inhibitor SB431542) block induction of the posterior primitive streak and prevent expression of *HOXB1* and *MESP1*.

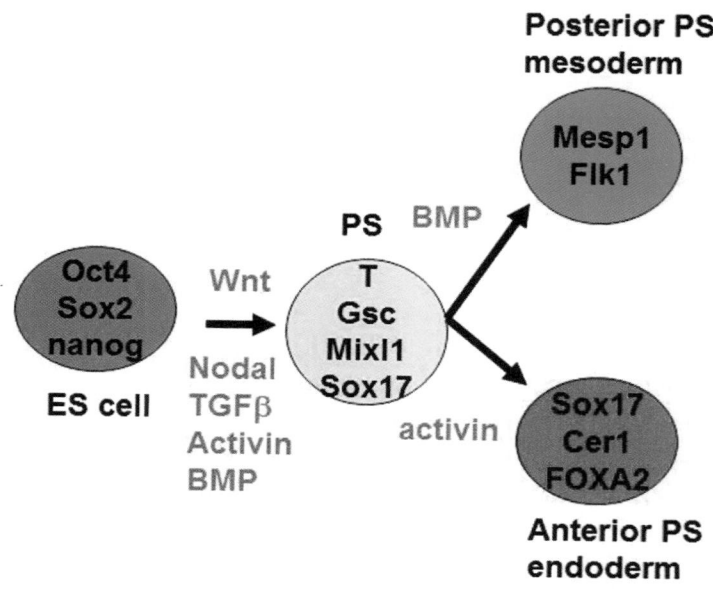

BMP4 also reveals a population of cells expressing *FLK1*, an early marker of the hematopoietic fate. *Brachyury*- and *FLK1*-positive cells can be FACS-sorted and induced to form hematopoietic colonies within cell aggregate in the presence of VEGF. WnT is crucial in this developmental pathway as DKK blocks erythroïd differentiation, a cell population expressing *CD41*, *GATA1*, and *SCL* (*stem cell leukemia*) [13].

ES cells can recapitulate the first stages of cardiogenesis, including specification and differentiation toward a cardiac phenotype. Fifteen years ago, a pioneering paper [14] reported that mouse ES cells were able to differentiate in vitro toward all cardiac cell types (i.e., atrial, ventricular, conduction, or pacemaker cells). Human ES cells also feature

12

the potential to differentiate into cardiac myocytes [15]. The spontaneous cardiac differentiation can be improved using BMP2 (bone morphogenetic protein) or TGFbβ , two members of the TGFb β superfamily secreted by the endoderm, or after co-culture with endodermal cells [16].

Many transcription factors play a key role in cardiac cell specification. These factors belong to different gene families, such as the *homeoproteins Nkx*, the *MADS-box family (Myocyte Enhancer Factor-2, Mef2)*, the *zinc finger factors (GATA family)*, and the *T-box family (Tbx)*. These factors are integrated in a transcriptional network. The cardiac transcription factors are more specifically Nkx2.5, Mef2c (Myocyte Enhancer factor), GATA4, GATA5, GATA6, myocardin, Tbx5, and Tbx20. These factors work in concert to turn on the transcriptional activity of promoters of constitutive cardiac genes [17].

3.2.3 Differentiation toward endoderm (liver, pancreatic)

Activin used at high dose (100 ng/ml) together with Nodal are potent inducers of the definitive endoderm. CXCR4, an antigen present at the surface of definitive endodermal cells, is then expressed. Stimulation of both mouse and human ES cells with these factors triggers expression of *FOX2 (Forkhead 2)* and the *homeobox* gene *MIXL1* by day five and then *SOX17, HEX, HNF4 (hepatocyte nuclear factor 4)* by day six, and *PDX1 (pancreatic and duodenal homeobox gene-1)*, a pancreatic marker, by day seven. *FOXA2* and *SOX17* factors are more reminiscent of a bipotential stage of development known as mesendoderm, while *MIXL1* is a marker of the posterior primitive streak already committed to mesoderm. The presence of serum favors the posterior primitive streak and the mesoderm at the expense of the definitive endoderm.

To further direct the differentiation of the definitive endo-derm toward specific lineages, several growth factors are needed. FGF and retinoic acid, associated with a blockade of both SHH and WnT pathways, leads to a pancreatic fate of definitive endodermal cells. Nevertheless, to generate en-docrine cells secreting insulin, further stimulation of cells with HGF (hepatic growth factor), IGF1 (insulin growth factor 1) and glucagon-like peptide 1 analog exendin-4 are required together with an inhibitor of the Notch pathway [18]. These cells express at a stage of progenitors *PDX1*, *SOX9*, *HES1* (*hairy and enhancer of split 1*), *RBPJ* (*Retinol Binding protein J*) and then *PDX1*, *HLXB9* (*homeo box HB9*), *PAX4*, *PAX6*, *ISL1*, *NKX2.2* and *NKX6.1* when they are capable of secreting insulin.

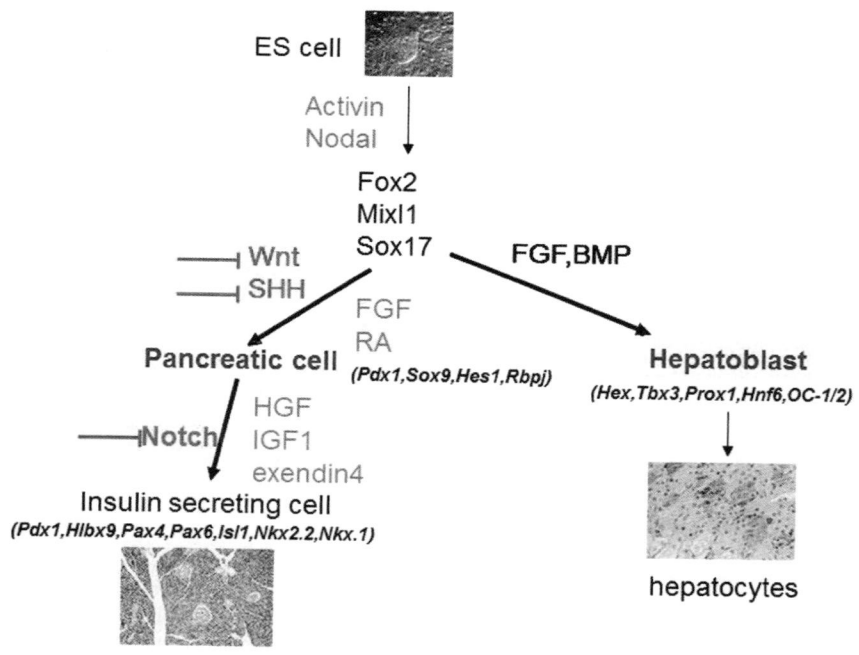

To direct differentiation toward a liver fate, both FGF and BMP must stimulate the embryonic endoderm. The responsive cells, or hepatoblasts, express *HEX*, *TBX3*, *PROX1* (*Homeobox prospero-like protein*), *HNF6*, and *OC-1/2* (*Osteocalcin*). [19]

4. Research on ES Cells

4.1 Cognitive research in biology of development

While early on ES cells were considered a homogeneous cell population, recent findings suggest rather that they are a heterogeneous population in a meta-stable state commuting from the inner cell mass to the epiblast stages of the embryo. This is thus a highly dynamic self-renewing cell population. To better understand this developmental stage of ES cells, one has to remember the embryonic development of mammals. In mouse embryo, at 3.5 dpc (days postcoitum), around 20 cells generate the inner cell mass. At 4.0, the cells from the ICM located at the border of the blastocoele differentiate into the primitive endoderm, while the cells in the ICM remain pluripotent. At this stage of development, called epiblast, cells remain pluripotent. At 5.5–6.0 dpc (i.e., gastrulation) the ICM/epiblast and then the early primitive streak (E6.5) proliferates and reaches about 600–700 cells and close to 10,000 cells at E7.5.

ES cells derived from the ICM are reminiscent of both the 4.0 dpc ICM and the 6.0 dpc epiblast. They thus model the early stages of differentiation. They express both pluripotent gene markers of the ICM such as *REX1*, *Stella*, and *TBX3*, as well as genes expressed in the epiblast such as *Brachyury* and *Goosecoid*, depending upon the cell colony. When differentiated within EBs, the pattern of expression of endodermal, mesodermal, and ectodermal genes is faithful to what is observed in the early embryo.

The ES cells thus provide a suitable cell model to answer questions in early embryology [20]. These include, but are not limited to:

- What is the origin and development of the primitive endoderm?

- How does the primitive endoderm give rise to the parietal (adjacent to the throphoectoderm in embryo) and visceral (adjacent to the ICM) endoderm?

- What role does ICM positioning play in the fate of the cell?

- What are the mechanisms driving the epiblast and primitive endoderm lineage specification in the developing blastocyst?

- To which extent does heterogeneity of ES cells reflect the heterogeneity of the ICM?

4.2 Applications in pharmaceutical and veterinary research

Pharmaceutical research can take advantage of human ES cells as a model of development to test potential teratogenic effects of molecules or to obtain differentiated functional cells to test toxicity of molecules.

The three-dimensional EB structures indeed offer a good model to screen molecules as to their safety. PCR tests to monitor expression of a pattern of genes specific to cell lineages can be used as a readout to evaluate the effect of a potentially curative molecule on early development of the embryo. Such an assay is easily scalable using high throughput screening (HTS) linked to real time PCR. Another approach can use ES cell lines genetically modified to expressed reporter genes (e.g., GFP) under the transcriptional control of a gene promoter specific to a cell lineage. The potential inhibitory effect of a molecule on a differentiation process can be monitored by epifluorescence microscopy.

EBs or purified derivatives from EBs can be used to screen potential toxicity of molecules. The beating activity of EBs or purified cardiomyocytes derived from EBs represent a good model to test cardiotoxicity of drugs. Both the pacing frequency and the amplitude of contraction can be monitored using electrode microarrays and video-edge monitors, respectively. Growth of axons from neurons derived from ES cells within EBs or in a co-culture cell model turns out to be easily monitored using appropriate cell tracking software. This also represents a suitable model to screen neuronal toxicity of drugs.

Another application of ES cells is the production of recombinant vaccines. A chicken ES (EB66) cell line has been derived and maintains a genetic stability and a diploid karyotype. The cell line does not feature any adventitious agents (ALV, avian viruses) or reverse transcriptase activity. As with any ES cell line, the EB66 line's proliferation is unlimited, and it can be cultured at high cell densities as suspension cells (>20 million cells/ml) under animal serum–free conditions. More than twenty-five human and veterinary vaccines can be produced by the EB66 cell line. This limits the use of chicken eggs and avoids the potential harmful effect of viruses on eggs.

5. To Learn More

5.1 References

1. Ramalho-Santos M. and Willenbring H. 2007. On the origin of the term "stem cell." *Cell Stem Cell* 1:35–8.

2. Evans M. J. and Kaufman M. H. 1981. Establishment in culture of pluripotential cells from mouse embryos. *Nature* 292:154–6.

3. Martin G. R. 1981. Isolation of a pluripotent cell line from early mouse embryos cultured in medium conditioned by teratocarcinoma stem cells. *Proc Natl Acad Sci USA* 78:7634–8.

4. Reubinoff B. E., Pera M. F., Fong C. Y., Trounson A., and Bongso A. 2000. Embryonic stem cell lines from human blastocysts: somatic differentiation in vitro. *Nat Biotechnol* 18:399–404.

5. Thomson J. A., Itskovitz-Eldor J., Shapiro S. S., Waknitz M. A., Swiergiel J. J., Marshall V. S., Jones J. M. 1998. Embryonic stem cell lines derived from human blastocysts. *Science* 282:1145–7.

6. Thomson J. A., Kalishman J., Golos T. G., Durning M., Harris C. P., Becker R. A., and Hearn J. P. 1995. Isolation of a primate embryonic stem cell line. *Proc Natl Acad Sci USA* 92:7844–8.

7. Doetschman T. C., Eistetter H., Katz M., Schmidt W., and Kemler R. 1985. The in vitro development of blastocyst-derived embryonic stem cell lines: formation of visceral yolk sac, blood islands and myocardium. *J Embryol Exp Morphol* 87:27–45.

8.   Schwartz P. H., Brick D. J., Stover A. E., Loring J. F., and Muller F. J. 2008. Differentiation of neural lineage cells from human pluripotent stem cells. *Methods* 45:142–58. Epub 2008 May 29.

9.   Erceg S., Ronaghi M., and Stojkovic M. 2009. Human embryonic stem cell differentiation toward regional specific neural precursors. *Stem Cells* 27:78–87.

10. Bagutti C., Wobus A. M., Fassler R., and Watt F. M. 1996. Differentiation of embryonal stem cells into keratinocytes: comparison of wild-type and beta 1 integrin-deficient cells. *Dev Biol* 179:184–96.

11. Coraux C., Hilmi C., Rouleau M., Spadafora A., Hinnrasky J., Ortonne J. P., Dani C., and Aberdam D. 2003. Reconstituted skin from murine embryonic stem cells. *Curr Biol* 13:849–53.

12. Aberdam E., Barak E., Rouleau M., de LaForest S., Berrih-Aknin S., Suter D. M., Krause K. H., Amit M., Itskovitz-Eldor J., and Aberdam D. 2008. A pure population of ectodermal cells derived from human embryonic stem cells. *Stem Cells* 26:440–4..

13. Nostro M. C., Cheng X., Keller G. M., and Gadue P. 2008. WnT, activin, and BMP signaling regulate distinct stages in the developmental pathway from embryonic stem cells to blood. *Cell Stem Cell* 2:60–71.

14. Maltsev V. A., Wobus A. M., Rohwedel J., Bader M., and Hescheler J. 1994. Cardiomyocytes differentiated in vitro from embryonic stem cells developmentally express cardiac-specific genes and ionic currents. *Circ Res* 75:233–44.

15. Kehat I., Amit M., Gepstein A., Huber I., Itskovitz-Eldor J., and Gepstein L. 2003. Development of cardiomyocytes from human ES cells. *Methods Enzymol* 365:461–73.

16. Puceat M. 2006. TGFbeta in the differentiation of embryonic stem cells. *Cardiovasc Res* 16:16.

17. Harvey R. P. 2002. Patterning the vertebrate heart. *Nat Rev Genet* 3:544–56.

18. Raikwar S. P. and Zavazava N. 2009. Insulin producing cells derived from embryonic stem cells: are we there yet? *J Cell Physiol* 218:256–63.

19. Zaret K. S. 2008. Genetic programming of liver and pancreas progenitors: lessons for stem-cell differentiation. *Nat Rev Genet* 9:329–40.

20. Tam P. P. and Loebel D. A. 2007. Gene function in mouse embryogenesis: get set for gastrulation. *Nat Rev Genet* 8:368–81.

21. Pearson H. 2006. Genetics: what is a gene? *Nature* 441:398–401.

5.2 Books

Anderson, Scott C., and Ann Kiessling. *Human Embryonic Stem Cells: An Introduction to the Science and Therapeutic Potential.* Boston, MA: Jones and Bartlett Publishers, 2003.

Chiu, Arlene, and Mahendra S. Rao, eds. *Human Embryonic Stem Cells.* Totowa, NJ: Humana Press, 2003.

Gardner, Richard L., David Gottlieb, and Daniel R. Marshak, eds. *Stem Cell Biology*. Cold Spring Harbor Monograph Series 40. Cold Spring Harbor Laboratories Press.

Sell, Stewart, ed. *Stem Cells Handbook*. Totowa, NJ: Humana Press, 2003.

Turksen, Kursad, ed. *Embryonic Stem Cells: Methods and Protocols*. Totowa, NJ: Humana Press, 2001.

Human Embryonic Stem Cells: The Practical Handbook. John Wiley and Sons Ltd, 2007

5.3 Websites

International Society for Stem Cell Research
http://www.isscr.org/science/index.htm

NIH Stem Cell Resources
http://stemcells.nih.gov/

Nature Journal: Insights Stem Cells
http://www.nature.com/nature/supplements/insights/stem_cells/

California Institute for Regenerative Medicine (CIRM): videos on stem cells including ES cells
http://www.cirm.ca.gov/StemCellBasicsVideos

Cold Spring Harbor Laboratory: a biology animation resource
http://www.dnalc.org/resources/animations/stemcells.html

6. Definitions

Potency: the capacity to differentiate into specialized cell types. This requires stem cells to be either totipotent, pluripotent, or multipotent to be able to give rise to any mature cell type, although unipotent progenitor cells are sometimes referred to as stem cells.

Totipotent: stem cells can differentiate into embryonic and extraembryonic cell types. Such cells can generate a complete, viable organism. These cells are produced from the fusion of an egg and a spermatozoid.

Pluripotent: stem cells are the descendants of totipotent cells and can differentiate into all cell types except extra-embryonic tissue (e.g., ES cells).

Multipotent: stem cells can differentiate into a limited number of cell types, usually those of a closely related family of cells.

Oligopotent: stem cells can differentiate into only a few cell types, such as lymphoid or myeloid stem cells.

Unipotent: cells can produce only one cell type, their own, but have the property of self-renewal which distinguishes them from non-stem cells (e.g., muscle satellite cells).

Teratocarcinoma: malignant cancer that arises in the testes or ovaries of adult.

Aneuploid: abnormal number of chromosomes; a type of chromosome abnormality.

Blastocyst: the spherical embryo at the time of implantation; that is, the attachment of the embryo to the uterine wall. The blastocyst consists of trophectoderm, blastocele, (cavity) and ICM.

Inner cell mass: pluripotent tissue inside the blastocyst that gives rise to the embryo proper and yolk sac tissue.

Trophectoderm: an extraembryonic, outside tissue layer of the early embryo that connects the embryo to the uterus and forms the placenta.

Morphogen: a signaling molecule governing the pattern of tissue development. It acts directly on cells to produce specific cellular responses dependent on concentration.

Gene: the basic hereditary unit in a living organism. All living organisms depend on gene expression. A modern working definition of a gene is "a locatable region of genomic sequence, corresponding to a unit of inheritance, which is associated with regulatory regions, transcribed regions, and/or other functional sequence regions [21].

Epigenetics: a word fusing *genetics* and *epigenesis* defined by Waddington in 1942; involves modifications of the activation of genes, without affecting the basic structure of DNA.

Core regulatory network: collection of DNA segments (genes) in a cell which interact each other (indirectly through their RNA and protein expression products) and with other substances in the cell, thereby governing the rates at which genes in the network are transcribed into mRNA. In general, each mRNA molecule goes on to make a specific protein (or set of proteins). Some proteins, though, serve only to activate other genes, and these are the transcription factors that are the main players in regulatory networks or cascades. By binding to the promoter region at the start of other genes they turn them on or off, initiating or stopping the production of another protein, and so on. Regulatory networks respond to the external environment. Sometimes a self-sustaining feedback loop ensures that a cell maintains its identity.

Teratoma: tumor containing a mixture of cells derived from all three embryonic germ layers.

Karyotype: test to identify and evaluate the size, shape, and number of chromosomes in a sample of cells.

Specification: a process by which cells acquire a specific but reversible fate.

Determination: a process by which cells acquire a specific and irreversible fate.

Patterning: a process by which cells acquire differential spatial pattern of cell fate.

Differentiation: a process by which cells express specific molecular markers that allow them to perform a specific function.

Fate: what a cell and its progeny will give rise to in a later stage of development; cells become progressively more restricted in their fate as development progresses.

Lineage: progeny of one mother cell.

Chapter 2

Isolation and Maintenance of Murine Embryonic Stem Cells

Junjiu Huang[a, b], Fang Wang[a,c], Xiaoying Ye[c], Jinmao Liu[c], David L. Keefe[b] and Lin Liu[b,c]

[a]School of Life Sciences, Sun Yat-sen University, Guangzhou 510275, China
[b]Department of Obstetrics and Gynecology, University of South Florida College of medicine, Tampa, FL 33647, USA
[c]College of Life Sciences, Nankai University, Tianjin 300071, China

For correspondence:
Lin Liu
E-mail: liutelom@yahoo.com

1. Brief History of Embryonic Stem Cell Research

Embryonic stem (ES) cells are derived from preimplantation embryos. ES cells are pluripotent such that they not only can maintain self-renewal and proliferation indefinitely in vitro, but also are able to differentiate into all three embryonic germ layers: ectoderm, endoderm, and mesoderm in vitro or in vivo. These include more than two hundred twenty cell types in the adult body. The mouse is an idea model species for mammalian research since it has a small body, short progenitive cycle, and extensive accumulated genetic research data. The history of nuclear transfer embryonic stem (ntES) cell research has been mostly based on mouse ntES cell research. In 1981, ES cells were first derived from mouse embryos by Martin Evans [64] and Matthew Kaufman [27]. After fourteen years, the first primate ES cell line was isolated by James Thomson at the University of Wisconsin–MadisonIn [94]. Three years later, they developed a technique to isolate and grow human ES cells in dishes [93]. This pluripotent, endlessly dividing cell has been hailed as a possible means for treating degenerative diseases, including diabetes, Parkinson's disease, Alzheimer's, spinal cord injury, heart failure, and bone marrow failure [50, 105]. *Science* magazine honored human stem cell research as 1999's "Breakthrough of the Year."

Yet, human ES cell research is always complicated by ethical, religionous, political, and legal problems [21, 101]. Another potential problem involves immunorejection, which could occur during cell transplantation and may negatively influence the application of ES cells in human cell therapy. The birth of the first cloned sheep, Dolly, in 1997 [104] and cloned mice the following year 1998 [102], provided the hope of overcoming the inmmunorejection by creating patient-specific ES cells using somatic cell nuclear transfer (SCNT). For a decade, researchers had successfully engaged nuclear transfer techniques to establish ntES cell lines in a range of animal species from various types of somatic cells. A research team led by Huizhen Sheng generated human ntES cells by nuclear transfer of human somatic nuclei into rabbit oocytes [12]. ntES cells exhibited properties similar to conventional human ES cells. But recent work suggested defects in reprogramming using recipient oocytes from different species and concern about the potential use of discordant animal oocyte sources to generate patient-specific stem cells [15]. Thus far, no authentic human ntES cells have been generated, even though autologous ntES cells generated through therapeutic cloning had been proposed as promising technology for future therapies [97]. Recently, scientists have successfully constructed human nuclear transfer embryos [29, 55, 83] based on several advanced techniques studies in mice, but the efficiency of creating nuclear transfer embryos and isolation ntES cells was too low to meet the need of clinical application. Recently, Dieter Egli and his colleagues determined that removal of the oocyte genome may be the primary cause of developmental failure after nuclear transfer. They made pluripotent human triploid ntES cells from blastocysts constructed by fusion of the somatic cell with an oocyte [72]. As this technique still needs to use oocytes to create new special embryos, which are considered new lives, it does not solve the ethical problems and creates more worry about human cloning.

At the same time, George Daley's group found that the first ntES cells claimed by Hwang et al. [39, 40] were mistakenly named human ntES cells. Actually, they were the first human parthenogenetic embryonic stem (pES) cells created in the world, and it has been suggested that pES cells could serve as a source of histocompatible tissues for transplantation [45]. pES cells are derived from parthenogenetic embryos which cannot develop to term since they lack paternal expression of imprinted genes and cannot develop a functional placenta for supporting fetal development [43, 73]. Mouse pES cells have been studied for more than two decades and show extensive differentiation potential in vitro and in vivo, but their true pluripotency was questioned, particularly considering low chimera production and deficiency in germline competence—a commonly used standard for testing genetic integrity and pluripotency of ES cells in rodents. pES cells were generated from parthenogenetic embryos developed from oocytes activated simply by ethanol, added with cytochalasin, which inhibits polar body extrusion and diploidizes chromosomes. As mechanisms of oocyte activation by sperm during fertilization became better understood, artificial methods for activation of oocytes have been improved to mimic sperm-induced oocyte activation, such that parthenogenetic embryos develop like normally fertilized embryos during preimplantation stages [5, 47, 59, 86, 95]. With improved methods for oocyte activation, pESC lines with higher quality have been isolated from mouse [41], rabbit [28, 103], buffalo [82], nonhuman primate [17, 100], and even human [57, 60, 78] cells. pES cells may serve as an additional source of histocompatible tissues for cell therapy [23, 44, 48, 49, 57, 78]. Notably, epigenetic status in pES cells is changed during isolation and culture in vitro of pES cells, leading to improved pluripotency of pES cells [33, 36, 41, 54]. Moreover, mouse parthonogenetic pups can be produced directly from pES cells by tetraploid embryo complementation (TEC), which contributes to placenta development [13]. This is exciting because full-term developmental potential pES cells suggest that pES cells can differentiate into all cell types and functional organs in a body. All these discoveries may lead to opportunities for providing various autologous stem cell sources for potential stem cell therapy in the future.

The first ever clinical trial using human ES cells received permission from the U.S. Food and Drug Administration in

January 2009. Ten patients with spinal cord injuries will receive repair cells derived from ES cells. This marks the beginning of what is potentially a new chapter in medical therapeutics. In addition, President Barack Obama had signed an order opening up federal funding for ES cell research, clearing the way for scientists to conduct broad experiments on human ES cells. As has been seen in past decades, research on murine ES cells will not only help with understanding fundamental mechanisms of developmental pluripotency and differentiation in cell biology, but also will continue to facilitate human ES cell research and clinical application.

2. Sources of Mouse Embryonic Stem Cells

Technology development in embryology, embryo isolation, culture, and micromanipulation has made it possible to construct different types of embryos such as parthenogenetic embryos, androgenetic embryos, and SCNT embryos, in addition to normally fertilized embryos. ES cells derived from these different sources of embryos have enriched our study of ES cells and enlarged our cell sources for cell therapy.

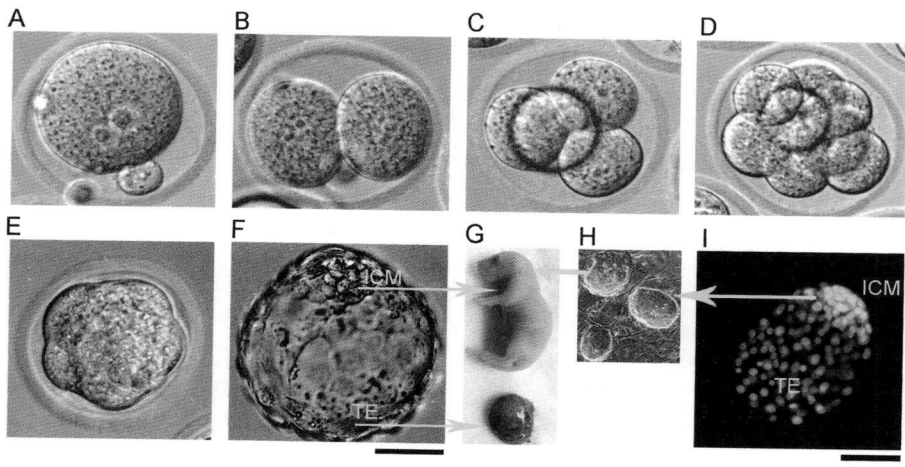

Figure 1. Mouse preimplantation embryo development cultured in KSO-M$_{AA}$ medium. Shown are early cleavage and developmental stages of mouse embryos: *A*, zygote; *B*, 2-cell; *C*, 4-cell; *D*, 8-cell; *E*, morula; and *F*, hatched blastocyst, inner cell mass cells (ICM), trophoblast cells (TE); *G*, a newborn pup developed from ICM and its placenta mostly developed from TE; *H*, embryonic stem cells at passage 2 isolated from ICM cells of B6C3F1 embryos; *I*, immunofluorescent staining showing *Oct4* positive for ICM cells (pink) and nuclei labeled with DAPI (blue). Yellow arrows show the origin of those cells or tissue. Bars, 50 μm.

2.1 Embryonic stem cells from fertilized embryos (fES cells)

fES cells are derived from normally fertilized preimplantation embryos. Early mouse embryo development can be achieved by culture in vitro in potassium simplex optimized media (KSO-M$_{AA}$) (figure 1 *A–F*). Blastocysts developed in vitro or blastocysts at 3.5 dpc developed in vivo flushed from the uterus horn can be used for isolation of ES cells. Embryos at 1-cell stage to 8-cell stage prior to compact stage are totipotent since any blastomeres of these embryonic cells can differentiate into embryonic and extraembryonic cell types. Such cells can construct a complete and viable organism including placenta tissues (figure 1 G). ES cells are usually derived from ICMs (figure 1 *H* and *I*), which can develop to all embryonic cell types of fetus and some extraembryonic cell types of placenta (figure 1 G and *H*). Isolation of fES cells from ICM cells of blastocysts is a routine method to produce ESC lines. However, some concern remains about the ethical issue of the destruction of embryos, which may be considered a life. To alleviate ethical concerns, modified methods for creating new ES cell lines that avoid destruction of embryos are being developed:

1. ES cell lines could be generated from single blastomere of mouse and human 8-cell stage embryos (figure 1 *D*) by using a single-cell biopsy similar to that routinely used in preimplantation genetic diagnosis (PGD) technique, without affecting developmental potential of micromanipulated embryos [16, 46]. But more than one blastomere might be needed and the efficiency of this technigue was low (about 2%).

2. Poor-quality embryos that most likely will not be able to develop to lives after fertility treatment are not used for embryo transfer but often discarded in IVF clinics. These poor-quality embryos might be used to generate ES cell lines [51, 52, 112]. Since these embryos are of poor quality, the efficiency of this method again is very low. These low quality embryos may have genome or chromosome defects. Their long-term effects on potential therapy require further investigation.

3. Trophectoderm cells mostly contribute to placenta, and embryos with defects in trophectoderm cells do not survive. Cdx2 is required for correct cell fate specification and differentiation of trophectoderm [84]. Cdx2 deficient embryos fail to develop to full term. However, pluripotent ntES cells can be generated from cloned Cdx2-deficient blastocysts using nuclear transfer [66]. The efficiency of this method is also very low, even in

31

mouse models. Additionally, safety issues regarding gene modification, which may have currently unknown risks, need to be addressed for cell therapy.

4. ES-like cells were achieved by fusing somatic cells with ES cells [19, 87]. These ES cells show tetraploidy, and it will be challenging to remove the genome of the original ES cells to turn the tetraploid to normal diploid ES cells.

2.2 Nuclear transfer embryonic stem (ntES) cells

Nuclear transfer can be used to generate autologous ES cell lines from a patient's own somatic cells. By nuclear transfer, a nucleus from somatic cells of the patient is transferred into an egg from which the chromosomal nucleus has been removed. The reconstituted activated eggs develop to blastocysts from which ntES cells are isolated. Maternal factors in eggs reprogram the genome from the somatic cell to turn it back to totipotent status.

2.3 Parthenogenetic embryonic stem (pES) cells and androgenetic embryonic stem (aES) cells

pES cells are isolated from parthenogenetic/gynogenetic embryos (two maternal chromosome sets), and aES cells from androgenetic embryos (two paternal chromosome sets).

1. Androgenetic embryos are usually generated by nuclear transplantation by micromanipulators [2]. Some reports have shown that aES cells are

pluripotent and can develop to many types of cells [24, 25, 61, 62, 92].

2. Diploid parthenogenetic embryos can be obtained by different methods [8]. The most common method uses chemicals to activate oocytes and obtain parthenogenetic embryos [8, 44]. Alternatively, parthenogenetic embryos containing genome from non-growing and fully grown oocytes are produced by germinal vesicle (GV) transfer [58, 81].

3. Molecular Mechanism of Pluripotency Regulation

Although long-term culture and maintenance of ES cells require feeder cells with the addition of mouse leukemia inhibitory factor (mLIF), mouse ES cells themselves have ground state for self-replication that does not require extrinsic instruction [110]. Many factors regulate pluripotency of ES cells, including transcription factors, epigenetic modification factors, microRNA, and many signal pathways. A few examples are listed below.

3.1 Transcription factors

OCT4 (Octamer-4) is a commonly used synonym for *POU5F1* (*POU class 5 homoeobox*). It is a homeodomain transcription factor of the POU family, which is critically involved in the self-renewal of ES cells [75]. *OCT4* is frequently used as the most important marker for ES cells. Yet, *OCT4* expression must be exactly regulated—higher or lower expression leads to differentiation of ES cells [52]. *SOX2*, an HMG box transcription factor, is essential for stabilizing ES cells in a pluripotent state by maintaining the requisite level of *OCT4* expression [65].

NANOG is a homeobox transcription factor, which plays a crucial role in the second embryonic cell fate specification, following the formation of the blastocyst [9]. *NANOG* expression in ICM cells of blastocyst prevents their differentiation into extraembryonic endoderm and trophectoderm. It is a divergent homeodomain protein that directs propagation of undifferentiated ES cells [11].

TCF3, a DNA-binding effector of Wnt signaling, is an integral component of the core regulatory circuitry of ES cells. It regulates pluripotency and self-renewal of ES cells by the transcriptional control of multiple lineage pathways, including reduced Nanog promoter activity and Nanog levels in mouse ES cells [18, 90, 107]. Further research shows that Tcf3 is a repressor, competing with the activity of *OCT4*, *SOX2*, and *NANOG*, and β–catenin inhibits its repression by direct interaction with *TCF3*. [106, 108]

Ronin possesses a THAP domain, which is associated with sequence-specific DNA binding and epigenetic silencing of gene expression. Ronin is essential for embryogenesis and the pluripotency of mouse ES cells by binding directly to a key transcriptional regulator protein host cell factor-1 (HCF-1) [22].

3.2 Epigenetic modification

DNA is tightly assembled with histone proteins and many other chromosomal proteins into a structure called chromatin. Changes in chromatin structure are affected by epigenetic modifications, such as histone modifications (including methylation, acetylation, phosphorylation, and ubiquitination) and DNA methylation inducing by non-histone, DNA-binding proteins. Epigenetic modification plays a key role in the regulation of gene transcription in ES cells [1, 30]. Polycomb group (PcG) proteins are transcriptional repressors that help to maintain epigenetic modification of chromatin structure through cell divisions [79] and also repress developmental regulators in ES cells [71]. The chromatin remodelling factor, Chd1, is required to maintain the open chromatin of pluripotent mouse embryonic stem cells [32].

3.3 MicroRNAs

microRNAs (miRNAs) are single-stranded RNA molecules of 21–23 nucleotides in length, which regulate gene expression. miRNAs play an important role in regulating ES cells self-renewal and differentiation [63]. Their main functions are repressing the translation of selected mRNAs in ES cells, such as let-7 suppress ES cells self-renewal [67] and miR-290-295 cluster maintains ES cells pluripotency [31, 37].

3.4 Signaling pathway

In LIF-STAT3 signaling pathway, leukemia inhibitory factor (LIF) is routinely used to maintain mouse ES cell culture at undifferentiated state. LIF binds to the specific LIF receptor (LIFR-α) which forms a heterodimer with a specific subunit common to all family members of the receptors, the *GP130* signal transducing subunit. This leads to activation of the JAK/STAT (Janus kinase/signal transducer and activator of transcription) and MAPK (mitogen activated protein kinase) cascades to regulate mouse ES cells [53, 69]. In mouse ES cells, *Klf4* is mainly activated by the JAK/STAT pathway and preferentially activates *Sox2*, whereas *Tbx3* is preferentially regulated by the phosphatidylinositol-3-OH kinase-Akt and MAPK pathways and predominantly stimulates *Nanog* [71].

In BMP pathway, bone morphogenetic proteins (BMPs) are the members of TGF-beta superfamily of secreted signaling molecules, which have important functions in many biological contexts [99]. BMP4, an important member of BMP signal pathway, regulates self-renewal of ES cells by means of induction of inhibitor of differentiation (Id) proteins and inhibition of both extracellular receptor kinase (ERK) and p38 MAPK pathways [76, 109].

Wnt signaling pathway involves a large number of proteins that can regulate the production of Wnt signaling molecules, their interactions with receptors on target cells, and the physiological responses of target cells that results from the exposure of cells to the extracellular Wnt ligands [35]. GSK3 (*glycogen synthase kinase-3*) inhibitor, *6-bromoindirubin-30-oxime* (*BIO*), which activates the Wnt/βcatenin pathway, also promotes self-renewal of ES cells [80, 96]. The mechanism is such that β–catenin can interact with Tcf3 and Tcf1 transcription factors to regulate gene expression in ES cells [108].

3.5 Core regulatory network for pluripotency

ES cell self-renewal and pluripotency are regulated by a network with a handful of factors, and any of these network components could shift a cell's equilibrium between self-renewal and differentiation. A trio of transcription factors is widely recognized as essential to ESC pluripotent state: *Oct4*, *Nanog*, and *Sox2*. These proteins interact closely with each other and often bind the same or nearby sites on DNA. *Nanog*, *Oct4*, and *Sox2* co-occupy at least 353 target genes, which include many homeodomain-containing transcription factors. They create a regulatory circuitry composed of autoregulation and feed-forward loops in ES cells [6, 74]. *Nanog* and *Oct4* associate with unique repressor complexes, which named Hdac1/2- and Mta1/2-containing complex *NANOG* and

OCT4 associated deacetylase (NODE), on their target genes to control ESC fate [56]. NODE complex contains histone deacetylase (HDAC) activity that seemed to be comparable to NuRD, which is comprised of the histone deacetylases *HDAC1* and *HDAC2* and usually associates with transcriptional repression [20, 56]. Nanog-induced ES cell self-renewal is LIF/STAT3 pathway independent. Notably, overexpression of Nanog supports LIF-independent self-renewal of mouse ES cells in the absence of *Klf4* and *Tbx3* activity [34, 71]. Also, overexpression of *Nanog* allows ES cells to self-renew without obligatory BMP4 signals by binding to Smad1 and activating Id proteins [10, 85]. Nanog promoter activity and Nanog levels are inhibited by *Tcf3*, a component of the Wnt signal pathway and a dominant downstream effector in mouse ES cells[18, 90, 107]. *Oct4*, *Nanog*, and *Sox2* are the major transcription factors that maintain pluripotent state of ES cells. Ronin performs a different and parallel pathway to achieve the same result in ES cells by binding HCF-1 to modify histones and help regulate gene expression in ES cells [22]. No doubt, more exciting research will lead to discovery of new regulatory and signaling pathways and to achieve a complete picture for pluripotency of ES cells. In the following, we describe the basics of isolation and culture of mouse ES cells.

4. Isolation and Culture of Mouse Embryonic Stem Cells

4.1 Materials

1. Feeder cell layers: mouse embryonic fibroblasts (MEFs)

2. Mouse 3.5 dpc blastocysts: fertilized or parthenogenetic embryos

4.2 Chemicals and reagents

1. Pregnant mare serum gonadotropin (PMSG) (367222, Calbiochem)

2. Chorionic Gonadotropin (hCG) (CG-10, Sigma)

3. Hyaluronidase (H-6254, Sigma)

4. Light mineral oil (M-8410, Sigma)

5. Cytochalasin D (CCD) (C-8273, Sigma), highly toxic!

6. Strontium Chloride (S-0390, Sigma)

7. Mineral oil (M-8410, Sigma)

8. Dimethylsulfoxide (DMSO) (D-2650, Sigma)

9. Dulbecco's modified Eagle's medium (DMEM, high glucose) (11995, Invitrogen)

10. Knockout DMEM (10829-018, Invitrogen)

11. Fetal bovine serum (FBS) (SH30070.03E, Hyclone)

12. Knockout serum replacement (KSR) (10828-028, Invitrogen)

13. Mouse ESGRO leukemia inhibitory factor (mLIF, Chemicon International Inc.)

14. MEM non-essential amino acid solution, 100X (NEAA) (M-7145, Sigma)

15. L-Glutamine-200 mM (100X), liquid (25030-081, Invitrogen)

16. β-mercaptoethanol (M-7522, Sigma)

17. Penicillin/streptomycin (15140, Invitrogen)

18. Gelatin (G-1890, Sigma)

19. Mitomycin C (M-4287, Sigma), highly toxic!

20. Trypsin-EDTA solution: 0.25% trypsin in 1 mM EDTA (25200, Invitrogen)

21. Trypsin inhibitor solution (Ti) (T-6414, Sigma)

22. Water (W-1503, Sigma)

23. Phosphate Buffered Saline (PBS) (20012, Invitrogen)

4.3 Medium preparation

1. Mouse embryo culture medium: potassium simplex optimized medium (KSOM) supplemented with amino acids KSOM$_{AA}$ and HEPES (14 mM)-buffered KSOM (HKSOM) [3, 26, 58]

2. Parthenogenetic activation (PA) medium: Ca^{2+}-free KSOM medium supplemented with 10 mM Sr^{2+} (2M SrCl$_2$ stock in sigma water, aliquot in 5 µl in 1.5 ml tubes, 200X, stored at -20°C) and 2.5 µg/ml cytochalasin D (1µg/µl CCD stock in DMSO, aliquot 2.5µl in 1.5 ml tubes, stored at -20°C)

3. Gelatin: 0.1% gelatin in PBS

4. MEF medium: High glucose DMEM supplemented with 10% FBS, 1 mM L-glutamine and 50 IU/ml penicillin, 50 IU/ml streptomycin

5. ES medium: 80% knockout DMEM, 20% FBS or KSR (KSR for primary ES-like cells isolation only), supplemented with 1000 units /ml mLIF, 0.1 mM NEAA, 1 mM L-glutamine, 0.1 mM β-mercaptoethanol, 50 IU/ml penicillin, 50 IU/ml streptomycin

6. ES cell freezing medium: ES medium supplemented with 10% DMSO and 30% FBS

4.4 Equipment and tools

1. Stereoscopic Zoom Microscope (SMZ800, Nikon)

2. Inverted Microscope (ECLIPSE TE100-F, Nikon)

3. Inverted Fluorescence Microscope (Zeiss)

4. CO_2 Incubator (Model 3130, Forma Scientific, Inc., USA)

5. Vertical Laminar Flow Benches (CV-30/70, TELSTAR)

6. Microcentrifuges (Model 5415D Brinkmann, Eppendorf)

7. Hot Plate (9062, Labotect)

8. Water incubator

9. Heat-pulled transfer pipettes, with opening diameter of 100-200 um

10. Alcohol burner

11. 4-well plate (176740, Nunc)

12. Cryotube vials (368632, Nunc)

13. Petri dishes, 35mm, sterile, (351008, BD Falcon)

14. Needle (30G X 0.5" 305106, BD Falcon)

15. Syringe (309602, BD Falcon)

4.5 Procedures for isolation of ES cells from mouse embryos

4.5.1 Collection and culture of mouse preimplantation embryos

1. Female mice (strains dependent on experiment design) at 1–2 months of age are superovulated with 5 IU PMSG followed 46–48 h later by 5 IU hCG.

2. Fertilized embryos:

   1) Blastocysts developed from in vitro: Successfully mated females 20–21h after hCG injection are used for collecting fertilized embryos. Zygotes enclosed in cumulus masses are removed by pipetting after brief incubation in 0.03% hyaluronidase prepared in HKSOM , wash three times in HKSOM and $KSOM_{AA}$ and then incubate in 50-μl droplets of pre-equilibrate $KSOM_{AA}$, covered with embryo-tested mineral oil at 37°C in a humidified atmosphere of 6.5% CO_2 in air.

2) Blastocysts collected from in vivo: Dissect uterus horns from pregnant female mice 3.5 days after mating. Flush 3.5 dpc embryos from uterine horns using prewarmed HKSOM medium. Blastocysts are picked up with polished transfer pipettes and cultured in 50-µl droplets of pre-equilibrate KSOMAA, covered with embryo-tested mineral oil at 37°C in a humidified atmosphere of 6.5% CO2 in air.

3. Parthenogenetic embryos: Oocytes enclosed in cumulus masses are collected from oviduct ampullae 14 h after hCG injection without mating. Cumulus cells are removed by pipetting after brief incubation in 0.03% hyaluronidase prepared in HKSOM. Oocytes are parthenogenetically activated by $SrCl_2$ and cytochalasin D in PA medium for 4 hours in incubator, and then washed more than three times in 50-µl droplets of $KSOM_{AA}$ medium and cultured in 50-µl droplets of pre-equilibrate $KSOM_{AA}$ covered with embryo-tested mineral oil in humidified atmospheres of 6.5% CO_2 to achieve diploid parthenogenetic blastocysts.

4. All in vitro manipulations are carried out at 37°C on a hot plate, chambers or incubators. Embryos are examined for cleavage to 2-cells at day 1–1.5 (20–24 h in culture, figure 1 B), and morula at day 2.5 for hybrid mouse strains (48-72 h in culture, figure 1 E). Blastocysts are formed at day 3.5 (96 h in culture, figure 1 F and figure 2 A). Embryos from inbred mice develop slower by 0.5 day.

4.5.2 Isolation of primary ES cell clones

1. D1: Collect zygotes and PA 1-cell embryos and culture them in $KSOM_{AA}$ until they develop to blastocysts stage at 3.5 days (figure 2 A).

2. D2: Thaw MEFs to 4-well dishes treated with 0.1% gelatin for 20 min at room temperature (RT). MEFs in each well are cultured in 850 µl MEF medium.

3. D3: Change with 850 µl MEF medium and keep MEFs culture one day more until MEFs reach 70–80% confluence.

4. D4: MEF cell layers are treated with 1µg/ml mitomycin C in MEF medium for 2.5 h, and then wash three times with PBS and culture in 500 µl MEF medium for an hour. Replace medium in feeder cell wells with freshly prepared KSR-ES medium (For primary ES-like cells isolation, we use KSR instead of FBS to make ES medium). Rinse blastocysts three times through KSO-M$_{AA}$, then ES medium. Transfer three to five blastocysts (cultured in vitro or flushed in vivo.) per well using heat-polished transfer pipettes preferably in the center of the well. A steromicroscope placed in the vertical laminar flow or on a bench with clean air flow is used for the procedures under sterile conditions.

5. D5: Do not touch and move wells, and keep culture for one day more to allow blastocysts to hatch from zona pellucidae (ZP), attach, and form outgrowths.

6. D6–D12: Usually, more than half of blastocysts hatch and adhere to the feeder cells on the second day after seeding and ICM cells grow well in KSR-ES medium (figure 2 B). Half of the medium is changed daily. Approximately 8–10 days after seeding, ICM outgrowths appear larger with typical colony shape (figure 2 C).

7. D11–13: Outgrowths are mechanically removed by polished transfer pipettes and digested with 0.25% trypsin-ETDA into small clumps or single cell, and then stopped digesting with trypsin inhibitor (Ti) and

reseeded on fresh feeder cells treated with mitomy-cin C already. If outgrowths' cell masses are hard to be digested by trypsin use two 30G needles to cut them into small clumps mechanically in KSR-ES medi-um and reseed on fresh MEFs. Usually, use these two methods together to passage outgrowths as quickly as possible. Stable ES-like cell lines are routinely ob-tained after one or two passages (figure 2 D).

8. Afterward, ES-like cell lines (figure 2 D) are pas-saged and cultured in FBS-ES medium instead of KSR-ES medium following brief digestion with 0.25% trypsin-ETDA.

9. When ES cells reach near confluence in a well, they should be passaged at 1:8 into new feeder wells. Slow growing colonies can be trypsinized and re-plated on the same well to prevent from differentiation.

10. Stable ES cell lines are established after several pas-sages (figure 2 E).

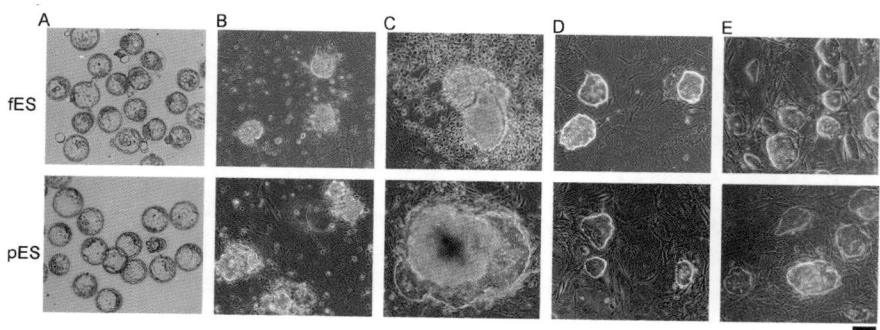

Figure 2. Isolation of primary ES cell colonies from fertilized C57/Bl6 embryos and parthenogenetic C57/Bl6 embryos. A, Fertilized embryos and parthenogenetic embryos cultured in KSOM$_{AA}$ medium for 96 h. B, ICM outgrowths after seeding on MEF cells at day 2. C, ICM outgrowths (primary ES colonies) after seeding on MEF cells at day 8. D, First pri-

mary passage of ESC clones. *E*, ESC lines at passage 9. fES cells, fertilization embryonic stem cells; pES cells, parthenogenetic embryonic stem cells. Bar, 100 μm.

4.5.3 ES cell freezing and storage

1. Remove ES medium and wash ES cells once with PBS, when ES cells grow near confluence (figure 2 *E*). Add one drop of preheated 0.25% trypsin-ETDA into each well of 4-well for a minute and then use 800 μl ES medium to block digesting.

2. Pipette ES cells mass into single cell and transfer them to a sterile 1.5 ml tube.

3. Pellet the cells at 800-1200 rpm for 6 min at room temperature.

4. Remove the supernatant and resuspend cells gently in 800 μl freezing medium ; mix them by pipetting several times.

5. Quickly aliquot 800 μl of the cell suspension into two labeled cryovials (400 μl each) and put them in a −80°C freezer overnight, then transfer cryovials into a liquid nitrogen tank for long-term storage.

4.5.4 ES cell thawing

1. Remove the vial from liquid nitrogen and thaw cryovials by quickly warming it in an incubator with 37°C water.

2. When the ice crystals almost disappear, aseptically transfer the cell suspension into a 1.5 ml tube using a pipette filled with 500 μl of preheated ES medium to dilute DMSO.

3. Pellet the cells at 800–1200 rpm for 6 min and then re-suspend the pellet in fresh ES medium ; plate on a new feeder well and culture the thawed in incubator at 37°C.

4. Change whole medium to remove floating dead cells the next day and then change half ES medium daily. Cells should be ready for passaging in 2–3 days.

5. Characterization of Mouse Embryonic Stem Cells In Vitro

5.1 Molecular markers of ES cells

5.1.1 Materials

1. ES cells

2. MEFs

5.1.2 Chemicals and reagents

1. ES medium (see 4.3)

2. MEF medium (see 4.3)

3. Phosphate Buffered Saline (PBS) (20012, Invitrogen)

4. Trypsin-EDTA solution: 0.25% trypsin in 1 mM EDTA (25200, Invitrogen)

5. Gelatin (G-1890, Sigma)

6. Goat serum (G-9023, Sigma)

7. Bovine Serum Albumin (A-3059, Sigma)

8. Paraformaldehyde Powder 95% (158127, Sigma)

9. Potassium Chloride (P-5405, Sigma)

10. Methanol/MeOH (1412, Fisher)

11. Acetic acid (BP2401-500, Fisher)

12. Nocodazole (M-1404, Sigma)

13. Ethanol (E-7023, Sigma)

14. Triton X-100 (X-100, Sigma), highly toxic!

15. Blue Alkaline Phosphatase Substrate kit III (SK-5300, DAKO, Vector Labs)

16. Mouse anti Oct4, IgG (sc5279, Santa Cruz)

17. Rabbit polyclonal to Nanog (ab10626, Abcam)

18. Mouse anti SSAE-1 (MC-480, DSHB)

19. Texas red anti-mouse IgG (TI-2000, Vector)

20. FITC goat anti-rabbit IgG (554020, BD Biosciences Pharmingen)

21. Alexa Fluor 488 goat anti-mouse IGM (A-21042, Molecular Probe)

22. Vectashield Mounting Medium (H-1000, Vector Labs)

23. Vectashield Mounting Medium with DAPI (H-1200, Vector Labs)

5.1.3 Equipment and tools

1. Inverted Fluorescence Microscope (Zeiss)

2. 4-well plate (176740, Nunc)

3. Vertical Laminar Flow Benches (CV-30/70, TELSTAR)

4. CO_2 Incubator (Model 3130, Forma Scientific, Inc., USA)

5. Micro slides (2947, plain, Corning)

6. 35 mm petri dishes (351008, BD Falcon)

7. 60 mm petri dishes (351007, BD Falcon)

5.1.4 Morphology and immunofluorescent staining of ES cell markers

1. Thaw ES cells and culture in 4-well plates until they reach near confluence (figure 2 E). Take morphological pictures with DIC filter under inverted microscope (figure 3 A).

2. Fixation and blocking: Cells are washed twice in PBS and then fixed in freshly prepared 3.7% paraformaldehyde in PBS for 15 min on ice pack. Remove paraformaldehyde solution and treat cells with 0.1% Triton

X-100 in blocking solution (3% goat serum plus 0.1% BSA in PBS) for 30 min, wash three times 15 min each through rinses of fresh blocking solution, and leave in blocking solution for 1 h.

3. 1st Antibody: incubation at 4°C in the refrigerator overnight with primary antibody Oct4 diluted 1:50 (Nanog, 1:150 and SSEA 1, 1:100) in blocking solution.

4. 2nd Antibody: remove primary antibody and wash three times in blocking solution. Incubate for 1 h at room temperature with FITC-conjugated goat anti-mouse IgG diluted 1:200 (FITC goat anti-rabbit IgG, 1:200 or Alexa Fluor 488 goat anti-mouse IgM, 1:200 for SSEA 1) in blocking solution, wash three times in blocking solution.

5. Stain Nuclei: add 50 μl vectashield containing 0.5 μg/ml DAPI into each well.

6. Take immunofluorescent staining pictures of Oct4 (figure 3 B), Nanog (figure 3 C), and SSEA 1 (figure 3 D) of ES cells.

5.1.5 Alkaline phosphatase

1. Alkaline phosphatase (AP) staining is performed using the Vector blue kit from Vector Laboratories following its protocol.

2. ES cells clones appear blue under color camera or black under black and white camera (figure 3 E).

5.1.6 Chromosome spreads for karyotypes

1. Digest ES cells and MEF feeder cells with 0.25% trypsin-EDTA when ES cells reach near confluence in 35 mm dishes. Re-seed all cells in the same 35mm dish for 30 min until most of MEFs adhere to the bottom. Transfer suspension to a feeder-free 60 mm dish treated with 0.1% gelatin and culture for 1–2 days until ES cells reach near confluence.

2. Add 0.5 μg/ml nocodazole to new ES medium to arrest mitosis of ES cells for 1.5 h.

3. Digest ES cells with 0.25% trypsin for 2–3 min, add 5 ml ES medium to block digestion and then transfer into a 15 ml tube.

4. Pellet cells at 900 rpm for 8 min at room temperature (RT); discard supernatants as much as possible.

5. Hypotonic treatment in 10 ml 0.075M KCL for 25 min at RT.

6. Add 4 drops 3:1 fixative medium MeOH/Acetic acid (freshly prepared and cool down at –20°C), invert tubes gently and wait 3–5 min and then pellet cells at 900 rpm for 8 min; discard supernatants.

7. Add 3 ml fixative medium to pipette cell pellets more easily and then add 7ml fixative to final 10ml; invert tubes and keep at RT for 30 min.

8. Spin at 900 rpm for 8 min; discard supernatants.

9. Repeat steps 7 and 8 three times.

10. Add about 0.25–0.5ml (depending on number of cells obtained) fixative to resuspend cells.

11. Drop onto pre-cleaned slides (cleaned with 100% ethanol and stored in clean water at 4°C prior to use).

12. Dry slides at 30 degree angle on benchtop over-night at RT.

13. Mount 15 µl DAPI (0.5 µg/ml) vectashield on a slide and cover with a glass cover.

14. Take immunofluorescent staining pictures with a 63X oil lens (figure 3 F).

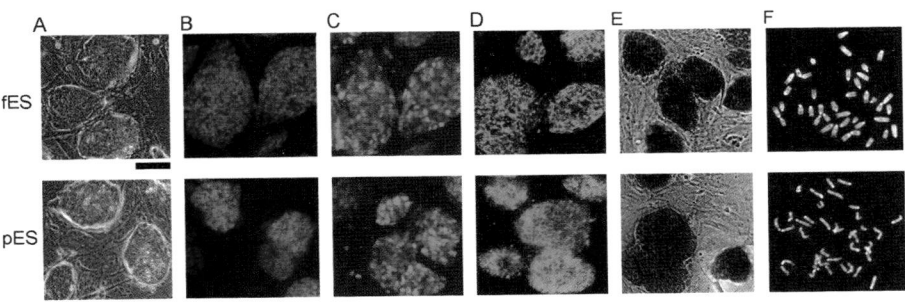

Figure 3. Molecular markers and karyotypes of C57/Bl6 fES cells and pES cells. A, morphology of ES cells under DIC microscopy at passage 9. B, immunofluorescent staining of Oct4 (red); C, Nanog (green); D, SSEA 1 (green) of ES cells at passage 9; nuclei were labeled with DAPI (blue). Note, feeder cells show blue staining but no specific staining for Oct4, Nanog, and SSEA1 around ES clones. E, expression of alkaline phosphatase in ES cells (AP, black) at passage 9. F, karyotypes of normal ES cells at passage 6. fES cells, fertilization embryonic stem cells; pES cells, parthenogenetic embryonic stem cells. Bar, 100 µm (A–E).

5.2 Induction of ES cell differentiation in vitro

5.2.1 Materials

1. ES cells

2. MEFs

5.2.2 Reagents and media

1. ES medium

2. ES medium without mLIF

3. Phosphate Buffered Saline (PBS) (20012, Invitrogen)

4. Trypsin-EDTA solution: 0.25% trypsin in 1 mM EDTA (25200, Invitrogen).

5. Gelatin (G-1890, Sigma)

6. Goat serum (G-9023, Sigma)

7. Bovine Serum Albumin (A-3059, Sigma)

8. Paraformaldehyde Powder 95% (158127, Sigma)

9. Triton X-100 (X-100, Sigma), highly toxic!

10. Alexa Fluor 568 goat anti-rabbit IgG (A-11011, Molecular Probe)

11. Alexa Fluor 488 goat anti-mouse IgG (A-11001, Molecular Probe)

12. Vectashield Mounting Medium (H-1000, Vector Labs)

13. Vectashield Mounting Medium with DAPI (H-1200, Vector Labs)

14. β-III-tubulin (CBL412, Chemicon)

15. Rabbit polyclonal to alpha smooth muscle actin (SMA) (ab5694-100, Abcam)

16. Polyclonal rabbit anti-human alpha 1-fetoprotein (AFP) (DAK-N150130, DAKO)

5.2.3 Equipment and tools

1. 6-well plate (140675, Nunc)

2. 100 mm petri dish (351005, BD Falcon)

3. CO_2 Incubator (Model 3130, Forma Scientific, Inc., USA)

4. Inverted Fluorescence Microscope (Zeiss)

5. Vertical Laminar Flow Benches (CV-30/70, TELSTAR)

5.2.4 Differentiation of ES cells into three germ layers in vitro

1. ES cells are trypsinized from 6-well plates and incubated for 50 min to remove feeder cells (figure 4 A).

2. Suspensions are centrifuged, and ES cells transferred to a 100 cm petri dish with ES medium without LIF for 2 days to form embryonic bodies (EBs) (figure 4 B) ; then large and well-shaped EBs are placed into 4-well plates, with three EBs per well, and cultured for 18 days.

3. EBs differentiation cells are processed for immuno-cytochemistry analysis. Immunostaining and micros-copy for characterization of three embryonic germ layers are performed as described above (see 5.1.4).

4. Cells stained with β-tubulin (green) are from ecto-derm layer (figure 4 C), alpha smooth muscle actin (SMA) (figure 4 D) from mesoderm layer, and alpha 1-fetoprotein (AFP) endoderm layer (figure 4 E).

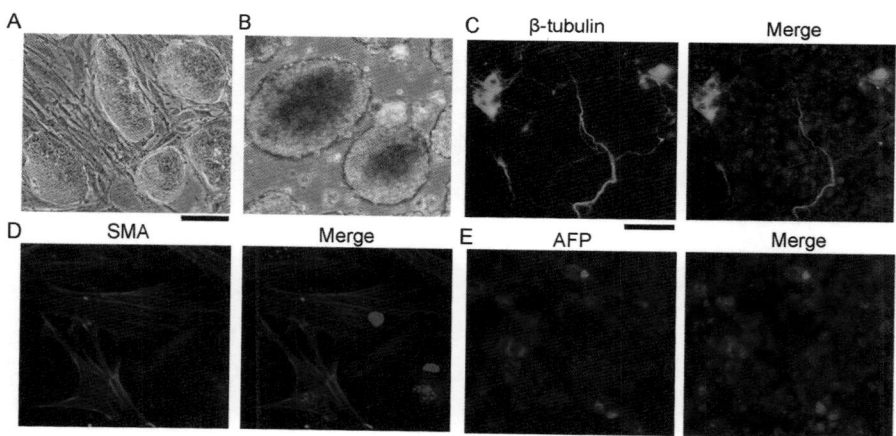

Figure 4. In vitro differentiations of C57/Bl6 fES cells into tissues of the three embryonic germ layers. A, C57 fES cells at passage 15. B, embry-onic bodies (EBs) derived from ES cells after culture in LIF-free ES me-dium for 2 days. C, immunofluorescence image of differentiation cells stained with ectoderm marker β-tubulin (green) antibodies, mesoderm marker alpha smooth muscle actin (SMA, red) antibody, and, D, en-doderm marker alpha 1-fetoprotein (AFP, red) antibody; E, nuclei are labeled with DAPI (blue). Bars, 100 μm (A and B) and 50 μm (C–E).

6. Pluripotency Test of Mouse Embryonic Stem Cells In Vivo

6.1 Materials

1. ES cells and MEFs

2. Immunodeficient nude mice, KM mice and Balb/c mice

3. 2.5 dpc pseudopregnant females

6.2 Reagents and media

1. ES medium (see 4.3)

2. $KSOM_{AA}$ and HKSOM (see 4.3)

3. Phosphate Buffered Saline (PBS) (20012, Invitrogen)

4. 0.25% Trypsin-EDTA solution (25200, Invitrogen)

5. Paraformaldehyde Powder 95% (158127, Sigma)

6. Polyvinylpyrrolidone (PVP) (P-0930, Sigma)

7. Mercury, highly toxic!

8. Proteinase (P-6911, Sigma)

9. D-Mannitol (M9647, Sigma)

10. $CaCl_2 .2H_2O$ (C7902, Sigma)

11. $MgCl_2$ (M8266, Sigma)

12. Bovine serum albumin (A9647, Sigma)

13. Hepes (H-6147, Sigma)

6.3 Equipment and tools

1. Petri dishes, 35mm, sterile (351008, BD Falcon)

2. Needle (30G, 305106, BD Falcon)

3. Syringe (309602, BD Falcon)

4. CO_2 Incubator (Model 3130, Forma Scientific, Inc., USA)

5. Microcentrifuge (Model 5415D, Brinkmann, Eppendorf)

6. Pipette puller (P-97, Sutter Instruments)

7. Microforge (MF-900, Narishige, Japan)

8. Automated inverted Leica DM IRE2 microscope and Eppendorf (Brinkmann) electric micromanipulators (Leica, German)

9. Air/Oil Microinjectors (Eppendorf CellTram, German)

10. Pizeo Micromanipulator (PMM-150FU, Prime Tech Ltd., Japan)

11. Borosilicate glass capillary tubing, length 10 cm. Standard wall outside diameter 1.0 mm; inside diam-

eter 0.58 mm. Thin wall outside diameter 1.0 mm and inside diameter 0.78 mm. (Warner Instrument Corp.)

12. Aggregation needle (DN-09, BLS, Hungary)

13. Electrode-chamber with 200-µm gap (Eppendorf or home made)

14. Multiporator (Eppendorf) or Electroporator (2001, BTX)

15. Stereoscopic Zoom Microscope (SMZ800, Nikon)

16. Heat-pulled transfer pipettes, with opening diameter of 100–200 µm.

17. Hot plate (9062, Labotect)

6.4 Teratoma Formation Test

1. Collect approximately 2×10^6 ES cells for testing and MEFs served as negative control and then dilute cells in 200 µl PBS solutions for injection immediately.

2. Inject subcutaneously into five-week-old immunodeficient nude mice to evaluate teratoma formation.

3. Four weeks after injection, mice with teratoma formation are sacrificed, and the resultant teratomas excised (Figure 5A), fixed in 4% paraformaldehyde, embedded in paraffin, and sectioned for histological examination.

4. Analysis of different types of tissues from different germ layers in teratoma, ectoderm (Figure 5B, a,

neurocyte), mesoderm (Figure 5B, b, muscle), and endoderm (Figure 5B, c, gland epithelium).

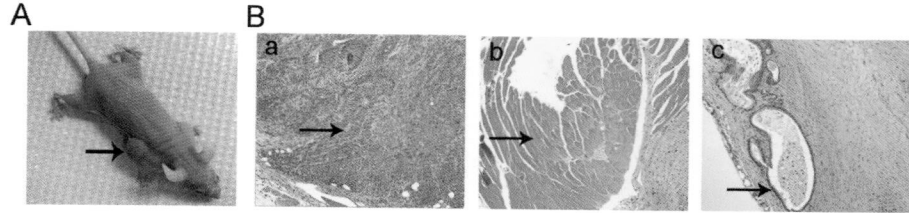

Figure 5. Tetratoma information of C57/Bl6 pES cells. A, teratoma formed from pES cells (right side) after injection of cells for five weeks; left side showed no teratoma from injection of fibroblast cells at passage 4 (MEFs P4). B, histological analysis of teratoma derived from ES cells. a, neuro-cyte; b, muscle; c, gland epithelium. Tissues shown by arrows.

6.5 Generation of chimera mice

6.5.1 Aggregation method

1. D1: Collect zygotes from albino females as described above (see 4.5.1). Choose ES cells with agouti coat for aggregation experiment. Thaw ES cells into 4-well plates with treated feeder cells at appropriate density, which could nearly reach confluence at D3.

2. D2 and D3: Culture embryos in KSOM$_{AA}$ medium for 2.5 days until they develop to 8-cell or morula stages (figure 1 D and E). Change ES medium daily.

3. D3: Aggregation.

   1) Place seven 50 µl KSOM$_{AA}$ microdrops into a 35 mm petri dish, covered with oil. Sterilize aggregation

needle by washing in 70% ethanol. Make six or more depressions in each microdrop and incubate at 37°C for more than an hour.

2) Treat with 2.5% protease in HKSOM for approximately 2–5 min at 37°C to remove the zonae pellucidae (ZP) of embryos°. Wash ZP-free embryos three times in both HKSOM and KSOM$_{AA}$ and then transfer one to each depression (figure 6 A).

3) Pick up some ES cell clones with a polished transfer pipette. Digest ES cell clones with 0.25% trypsin until they form mass of about 8–15 ES cells. Wash thesES cell mass in KSOM$_{AA}$ and place one cell mass (8–15 cells) into each depression, in close contact with the ZP-free embryo (figure 6 A).

4. Without shaking, put dishes gently back into the incubator and culture them together overnight. More details about aggregation method can be found in reference [91].

5. D4: Transfer aggregation blastocysts (figure 6 A) into uteri of 2.5 dpc pseudopregnant females.

6. Black (derivation from ES cells) and white chimera mice will be born 16–18 days after embryo transfer (figure 6 B).

6.5.2 Microinjection of ES cells into blastocysts

1. D1: Collect zygotes from albino females as described above (see 4.5.1).

2. D2 and D3: Choose ES cells from agouti mice for aggregation experiment. Thaw ES cells and culture in
   4-well plates with treated feeder cells at appropriate
   density, which could nearly reach confluence at D4.
   Culture embryos in $KSOM_{AA}$ medium for 3.5 days until
   they develop to blastocyst stage (figure 6 B). Change
   ES medium daily.

3. D4: Blastocysts Injection (BI).

   1) ES cells preparation: change ES medium 0.–2 hours
      before adding trypsin to the cells. Digest cells
      with 0.25% trypsin for 1–2 min and pipette them
      into single cell. Re-seed cells in the same well for
      30 min to remove most MEFs. Transfer ES cell sus-
      pension to a new 1.5 ml tube and spin at 900 rpm
      for 6 min at RT. Remove suspension and use 1 ml
      pre-cooled ES medium at 4°C added with 20 µl
      1 M Hepes/ml to resuspend ES cells. Put them at
      4°C for 30 min and remove 800 µl of upper part of
      medium to remove dead cells. Dilute ES cells with
      100–200 µl pre-cooled medium (dilution volume
      depends on how many ES cells are available ; this
      can be estimated under microscope before dilu-
      tion). Store ES cell suspension at 4°C and use them
      within three hours.

   2) Blastocysts: pick up expanded blastocysts, which
      have clear ICM cells but have not yethatched,
      for blastocyst injection. If no blastocysts are devel-
      oped from zygotes at D1, expanded blastocysts
      can be flushed from uteri of 3.5-day pregnant fe-
      males with HKSOM in the morning and cultured in
      $KSOM_{AA}$ until ES cells are ready to use.

3) Make holding pipettes and injection pipettes as described in the reference [58]. Set up micromanipulation and piezo micromanipulator for injection (figure 6 C). Transfer 10–30 blastocysts into HK-SOM drop for manipulation. Back-load 3 mm length mercury into an injection pipette. The blunt end of the pipette has a bore of approximately 15 μm in diameter. Push the mercury through the shoulder, nearly to the tip, to empty the air in the pipette ; then wash three times in both 10% polyvinylpyr-rolidone (PVP)-PBS solution and HKSOM medium. Suck roughly 1 mm HKSOM into the pipette to keep ES cells away from mercury and then suck dozens of ES cells into the pipette for injection. Maintain approximately 200 μm away from the pipette tip. Damage to cell membranes may occur if the cells are very close to the tip when the piezo pulses are applied. A blunt-end injection pipette is used to introduce about twelve ES cells into expanded blastocyst near the ICM cells (figure 6 D).

4. All embryos injected with ES cells are cultured in $KSOM_{AA}$ at 37°C in an atmosphere of 6.5% CO_2 for 2–4 h.

5. Re-expanded blastocysts are transferred to the uterine horns of 2.5 dpc pseudopregnant females by standard methods.

6. EGFP-fluorescent (green) chimera fetuses at 13.5 dpc can be found after injection of β-actin promoter-EGFP ES cells (figure 6 E), and black (from ES cells) and white chimera mice will be born after 16–18 days (figure 6 F).

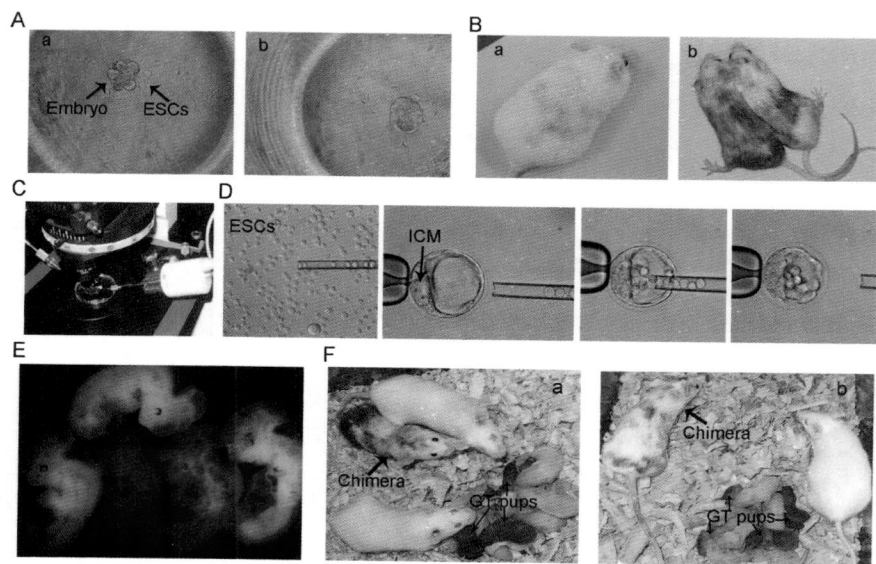

Figure 6. Methods of generating chimera mice. *A*, albino (KM) mouse embryo and B6C3F1 pES cells mass aggregation; *a*, ES cells in close contact with ZP-free 8-cell morula; *b*, after overnight incubation. pES cells, parthenogenetic embryonic stem cells. *B*, chimera mice produced by; *a*, KM embryo- B6C3F1 pES cells aggregation; *b*, fertilized Balb/c embryo-parthenogenetic B6D2F1 embryo aggregation. *C*, manipulation dish and micromanipulator setup for blastocyst injection. *D*, process of microinjection B6C3F1 ES cells into KM blastocysts. Arrow shows inner cell mass. *E*, EGFP fluorescent (green) chimera fetuses at 13.5 dpc after injection of β-actin promoter-EGFP ES cells. *F*, chimeras produced by; *a*, injection of B6C3F1 fES cells; *b*, pES cells, into Balb/c blastocysts by piezo, and their germline transmission (GT) pups. Arrows show chimera and GT pups with black coat produced from the chimera mated with an albino ICR mouse.

6.6 Production of ES cell–mice from ES cells

6.6.1 ES cell–mice produced by tetraploid embryo complementation (TEC) method

1. Tetraploid embryos.

    1) Collect zygotes from albino mice and culture them in $KSOM_{AA}$ until they reach 2-cell stage next day (figure 1 B and figure 7 A).

    2) Pick up 20–30 2-cell embryos and pipette them in 100 µl fusion medium (0.28 mM D-Mannitol, 0.1mM $CaCl_2$, 0.1mM $MgCl_2$ and 0.01% BSA in ultrapure water) through several drops to equilibrate.

    3) Transfer all embryos to electrode-chamber and place them between electrodes.

    4) Set up the parameters: AC 1.5 V, 5 s; DC 24 V (1.2 kV/cm), 50 µs, n=2; postfusion AC 0.0 V, 0s.

    5) Press start button: most of the embryos will correctly orient with cleavage plan parallel to the electrodes.

    6) Wash the embryos through several drops of HK-SOM and KOSMAA and then culture them in KOS-MAA. Most 2-cell embryos fuse after 30 min, but a few may take 1 h (figure 7 B). Remove non-fused embryos ; culture fused embryos until they develop to blastocysts (figure 7 C).

2. Procedure for tetraploid embryo injection is similar to blastocyst injection described above (see 6.5.2). For tetraploid embryos, inject about 20–25 ES cells into each blastocyst (figure 7 D).

3. Culture embryos injected with ES cells in KSOM$_{AA}$ at 37°C in an atmosphere of 6.5% CO$_2$ for 2–4 h (figure 7 E).

4. Transfer re-expanded blastocysts to the uterine horns of 2.5 dpc pseudopregnant females by standard methods.

5. Black ES cell mice derived from only ES cells will be born after 16–18 days (figure 7 F).

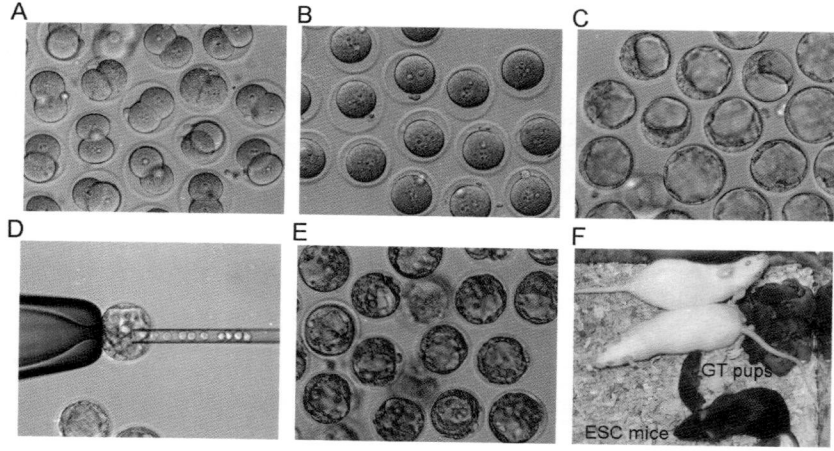

Figure 7. ES mice produced by tetraploid embryo complementation method. *A,* 2-cell embryos from KM mice cultured in KSOM$_{AA}$ medium. *B,* fused embryos at one cell stage. *C,* 3.5 dpc tetraploid blastocysts. *D,* injection of B6C3F1 ES cells into KM tetraploid blastocysts by piezo micromanipulator. *E,* tetraploid complementation blastocysts cultured in KSOM$_{AA}$ for 2-4 h after ES cells injection. *F,* ES cell male and its germline transmission (GT) pups following mating with albino CD1 females.

6.6.2 ES cell–mice produced by four- or eight-cell embryo injection

1. Fertilized embryos are cultured in $KSOM_{AA}$ until they develop to four- or eight- cell embryos at 2.0 or 2.5 dpc (figure 1 C and D).

2. For easy recognition of the phenotypes of the resultant fetuses and pups, eight- or four-cell embryos from albino KM mice with white eyes are used as recipients for supporting development of donor ES cells from agouti mice.

3. The setup for injection of ES cells into eight- or four-cell embryos before obvious compaction using piezo micromanipulator is similar to that for blastocyst injection (see 6.5.2). The injection pipette is positioned near the ZP above the perivitelline space between two blastomeres. The ZP is pierced by the pipette immediately after application of 2–3 electropulses in a very short time generated from the piezo manipulator, and then 8–10 ES cells are injected between the embryonic cells under the ZP, followed by slow withdrawal of the pipette (figure 8 A).

4. Embryos injected with ES cells are cultured in $KSOM_{AA}$ at 37°C at an atmosphere of 6.5% CO_2 until 3.5 dpc.

5. Healthy embryos (morula and blastocysts) that develop beyond the eight-cell stage are transferred into the uterine horns of pseudopregnant ICR females by the standard procedure. Injection of β-actin promoter-EGFP ES cells shows that ES cells always devote to ICM cells (figure 8 B)

6. Newborn mice are delivered by cesarean section from ES cells at embryonic day 19.5 following injection into eight-cell embryos (figure 8 C).

7. Agouti ES cell–mice derived only from ES cells are mated with albino KM females to exam their competence for germline transmission (figure 8 D).

8. Microsatellite analysis confirms that all types of tissues of ESC mice are from ES cells only, except for the placentas (figure 8 E). More details about this technique can be found in the reference [38].

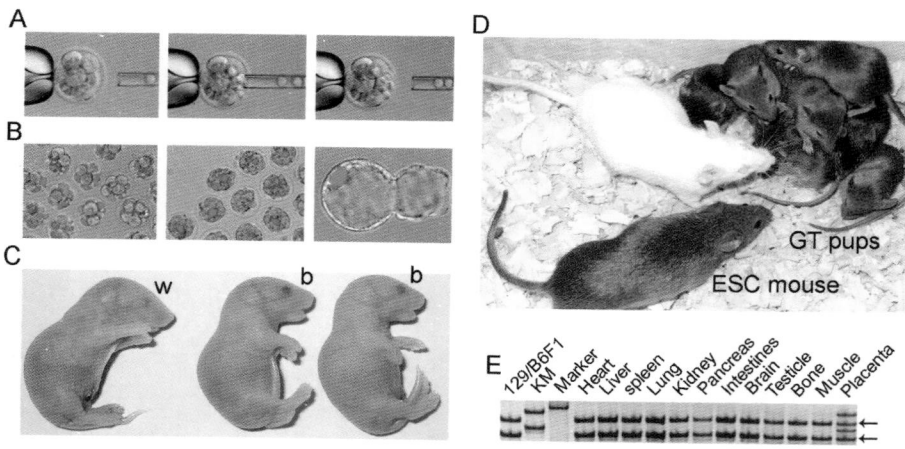

Figure 8. Pups produced by injection of ES cells into eight-cell recipient embryos. *A,* injection of ES cells into eight-cell embryos by piezo micromanipulation. *B,* injected β-actin promoter-EGFP ES cells (green) localized to ICM cells of blastocyst. *C,* newborn mice delivered by cesarean section from B6C3F1 ES cells at embryonic day 19.5 after injection into KM eight-cell embryos. Pups with white eyes are from host KM embryos, and pups with black eyes are from agouti B6C3F1 ESC-mice. *D,* ES cell–mouse from B6C3F1 ES cells injected into KM eight-cell embryos and its agouti pups, showing full germline transmission (GT). Albino KM females are used for mating. *E,* microsatellite analysis of DNA from different tissues of ESC mice. KM, Kunming mice. Arrows indicate bands from ES cells.

7. Future Stem Cell Research

Three decades have passed since the first mouse ES cell line was established in 1981. Based on the accumulated knowledge of ES cells, we know much more about the molecular mechanism of how ES cells regulate their pluripotency and how somatic cells can be reprogrammed by nuclear transfer and cell fusion with ES cells. In 2006, Japanese scientist Shinaya Yamanaka reported a breakthrough in the field of stem cell research—pluripotent stem cells had been induced (iPS cells) by transfection of four transcription factors into mouse somatic cells [89]. The following year two independent groups applied this technique to human material and achieved similar results [88, 111]. *Nature* magazine considered it the "most exciting scientific method," and *Science* magazine chose it as the "Breakthrough of the Year" [98]. In 2009, three different labs indicated that iPS cells had similar pluripotency to ES cells by producing viable adult iPS cell mice using the TEC method [4, 42, 113].

It seems that iPS cells, which have far fewer ethical concerns than fES cells, ntES cells, or pES cells, are the ideal autologous stem cell sources for cell therapy in the future. Yet, gene expression profiles of iPS cells are different from ES cells [14]. Also, in vivo experiments showed that secondary neurospheres (SNS) induced from various iPS cells exhibit different propensities to form teratoma [68]. These data suggest that iPS cells generated by the current technology are not identical to ES cells, and may be not safe. Thus, iPS cell research is still in its infancy, and much more needs to be learned from ES cells since ES cells are the current gold standard for stem cell research. Also, it is impossible to test true pluripotency of human iPS cells by making chimera or using the TEC method. Therefore, scientists must discover a gold standard for pluripotency examination for human ES cells and iPS cells. It is anticipated that new techniques and breakthroughs will come out quickly in the years to come, as stem cells recieve more and more attention in both basic and clinical research.

References

1. Atkinson, S., and Armstrong, L. 2008. Epigenetics in embryonic stem cells: regulation of pluripotency and differentiation. *Cell Tissue Res* 331:23–9.

2. Barton, S. C., Ferguson-Smith, A. C., Fundele, R., and Surani, M. A. 1991. Influence of paternally imprinted genes on development. *Development* 113:679–87.

3. Biggers, J. D., McGinnis, L. K., and Raffin, M. 2000. Amino acids and preimplantation development of the mouse in protein-free potassium simplex optimized medium. *Biol Reprod* 63:281–93.

4. Boland, M. J., Hazen, J. L., Nazor, K. L., Rodriguez, A. R., Gifford, W., Martin, G., Kupriyanov, S., and Baldwin, K. K. 2009. Adult mice generated from induced pluripotent stem cells. *Nature*. 461:91–94.

5. Bos-Mikich, A., Swann, K., and Whittingham, D. G. 1995. Calcium oscillations and protein synthesis inhibition synergistically activate mouse oocytes. *Mol Reprod Dev* 41:84–90.

6. Boyer, L. A., Lee, T. I., Cole, M. F., Johnstone, S. E., Levine, S. S., Zucker, J. P., Guenther, M. G., Kumar, R. M., Murray, H. L., Jenner, R. G., Gifford, D. K., Melton, D. A., Jaenisch, R., and Young, R. A. 2005. Core transcriptional regulatory circuitry in human embryonic stem cells. *Cell* 122:947–56.

7. Boyer, L. A., Plath, K., Zeitlinger, J., Brambrink, T., Medeiros, L. A., Lee, T. I., Levine, S. S., Wernig, M., Tajonar, A., Ray, M. K., Bell, G. W., Otte, A. P., Vidal, M., Gifford, D. K., Young, R. A., and Jaenisch, R. 2006. Polycomb complexes repress developmental regulators in murine embryonic stem cells. *Nature* 441:349–53.

8. Brevini, T. A., and Gandolfi, F. 2008. Parthenotes as a source of embryonic stem cells. *Cell ProLIF* 41 Suppl 1:20–30.

9. Cavaleri, F., and Scholer, H. R. 2003. Nanog: a new recruit to the embryonic stem cell orchestra. *Cell* 113:551–2.

10. Chambers, I. 2004. The molecular basis of pluripotency in mouse embryonic stem cells. *Cloning Stem Cells* 6:386–91.

11. Chambers, I., Colby, D., Robertson, M., Nichols, J., Lee, S., Tweedie, S., and Smith, A. 2003. Functional expression cloning of Nanog, a pluripotency sustaining factor in embryonic stem cells. *Cell* 113:643–55.

12. Chen, Y., He, Z. X., Liu, A., Wang, K., Mao, W. W., Chu, J. X., Lu, Y., Fang, Z. F., Shi, Y. T., Yang, Q. Z., Chen da, Y., Wang, M. K., Li, J. S., Huang, S. L., Kong, X. Y., Shi, Y. Z., Wang, Z. Q., Xia, J. H., Long, Z. G., Xue, Z. G., Ding, W. X., and Sheng, H. Z. 2003. Embryonic stem cells generated by nuclear transfer of human somatic nuclei into rabbit oocytes. *Cell Res* 13:251–63.

13. Chen, Z., Liu, Z., Huang, J., Amano, T., Li, C., Cao, S., Wu, C., Liu, B., Zhou, L., Carter, M. G., Keefe, D. L., Yang, X., and Liu, L. 2009. Birth of parthenote mice directly from parthenogenetic embryonic stem cells. *Stem Cells*. 27:2136–45

14. Chin, M. H., Mason, M. J., Xie, W., Volinia, S., Singer, M., Peterson, C., Ambartsumyan, G., Aimiuwu, O., Richter, L.,

Zhang, J., Khvorostov, I., Ott, V., Grunstein, M., Lavon, N., Benvenisty, N., Croce, C. M., Clark, A. T., Baxter, T., Pyle, A. D., Teitell, M. A., Pelegrini, M., Plath, K., and Lowry, W. E. 2009. Induced pluripotent stem cells and embryonic stem cells are distinguished by gene expression signatures. *Cell Stem Cell* 5:111–23.

15. Chung, Y., Bishop, C. E., Treff, N. R., Walker, S. J., Sandler, V. M., Becker, S., Klimanskaya, I., Wun, W. S., Dunn, R., Hall, R. M., Su, J., Lu, S. J., Maserati, M., Choi, Y. H., Scott, R., Atala, A., Dittman, R., and Lanza, R. 2009. Reprogramming of human somatic cells using human and animal oocytes. *Cloning Stem Cells.* 11:213–223

16. Chung, Y., Klimanskaya, I., Becker, S., Marh, J., Lu, S. J., Johnson, J., Meisner, L., and Lanza, R. 2006. Embryonic and extraembryonic stem cell lines derived from single mouse blastomeres. *Nature* 439:216–9.

17. Cibelli, J. B., Grant, K. A., Chapman, K. B., Cunniff, K., Worst, T., Green, H. L., Walker, S. J., Gutin, P. H., Vilner, L., Tabar, V., Dominko, T., Kane, J., Wettstein, P. J., Lanza, R. P., Studer, L., Vrana, K. E., and West, M. D. 2002. Parthenogenetic stem cells in nonhuman primates. *Science* 295:819.

18. Cole, M. F., Johnstone, S. E., Newman, J. J., Kagey, M. H., and Young, R. A. 2008. Tcf3 is an integral component of the core regulatory circuitry of embryonic stem cells. *Genes Dev* 22:746–55.

19. Cowan, C. A., Atienza, J., Melton, D. A., and Eggan, K. 2005. Nuclear reprogramming of somatic cells after fusion with human embryonic stem cells. *Science* 309:1369–73.

20. Crook, J. M., Dunn, N. R., and Colman, A. 2006. Repressed by a NuRD. *Nat Cell Biol* 8:212–4.

21. de Wert, G., and Mummery, C. 2003. Human embryonic stem cells: research, ethics and policy. *Hum Reprod* 18:672–82.

22. Dejosez, M., Krumenacker, J. S., Zitur, L. J., Passeri, M., Chu, L. F., Songyang, Z., Thomson, J. A., and Zwaka, T. P. 2008. Ronin is essential for embryogenesis and the pluripotency of mouse embryonic stem cells. *Cell* 133:1162–74.

23. Dighe, V., Clepper, L., Pedersen, D., Byrne, J., Ferguson, B., Gokhale, S., Penedo, M. C., Wolf, D., and Mitalipov, S. 2008. Heterozygous embryonic stem cell lines derived from non-human primate parthenotes. *Stem Cells* 26:756–66.

24. Dinger, T. C., Eckardt, S., Choi, S. W., Camarero, G., Kurosaka, S., Hornich, V., McLaughlin, K. J., and Muller, A. M. 2008. Androgenetic embryonic stem cells form neural progenitor cells in vivo and in vitro. *Stem Cells* 26:1474–83.

25. Eckardt, S., Leu, N. A., Bradley, H. L., Kato, H., Bunting, K. D., and McLaughlin, K. J. 2007. Hematopoietic reconstitution with androgenetic and gynogenetic stem cells. *Genes Dev* 21:409–19.

26. Erbach, G. T., Lawitts, J. A., Papaioannou, V. E., and Biggers, J. D. 1994. Differential growth of the mouse preimplantation embryo in chemically defined media. *Biol Reprod* 50:1027–33.

27. Evans, M. J., and Kaufman, M. H. 1981. Establishment in culture of pluripotential cells from mouse embryos. *Nature* 292:154–6.

28. Fang, Z. F., Gai, H., Huang, Y. Z., Li, S. G., Chen, X. J., Shi, J. J., Wu, L., Liu, A., Xu, P., and Sheng, H. Z. 2006. Rabbit embryonic stem cell lines derived from fertilized, parthenogenetic or somatic cell nuclear transfer embryos. *Exp Cell Res* 312: 3669–82.

29. French, A. J., Adams, C. A., Anderson, L. S., Kitchen, J. R., Hughes, M. R., and Wood, S. H. 2008. Development of human cloned blastocysts following somatic cell nuclear transfer with adult fibroblasts. *Stem Cells* 26: 485–93.

30. Gan, Q., Yoshida, T., McDonald, O. G., and Owens, G. K. 2007. Concise review: epigenetic mechanisms contribute to pluripotency and cell lineage determination of embryonic stem cells. *Stem Cells* 25: 2–9.

31. Gangaraju, V. K., and Lin, H. 2009. MicroRNAs: key regulators of stem cells. *Nat Rev Mol Cell Biol* 10:116–25.

32. Gaspar-Maia, A., Alajem, A., Polesso, F., Sridharan, R., Mason, M. J., Heidersbach, A., Ramalho-Santos, J., McManus, M. T., Plath, K., Meshorer, E., and Ramalho-Santos, M. 2009. Chd1 regulates open chromatin and pluripotency of embryonic stem cells. *Nature* 460:863–8.

33. Gong, S. P., Kim, H., Lee, E. J., Lee, S. T., Moon, S., Lee, H. J., and Lim, J. M. 2009. Change in gene expression of mouse embryonic stem cells derived from parthenogenetic activation. *Hum Reprod*. 24:805–814.

34. Hamazaki, T., Kehoe, S. M., Nakano, T., and Terada, N. 2006. The Grb2/Mek pathway represses Nanog in murine embryonic stem cells. *Mol Cell Biol* 26:7539–49.

35. Hlsken, J., and Behrens, J. 2000. The WnT signalling pathway. *J Cell Sci* 113 Pt 20:3545.

36. Horii, T., Kimura, M., Morita, S., Nagao, Y., and Hatada, I. 2008. Loss of genomic imprinting in mouse parthenogenetic embryonic stem cells. *Stem Cells* 26:79–88.

37. Houbaviy, H. B., Murray, M. F., and Sharp, P. A. 2003. Embryonic stem cell-specific MicroRNAs. *Dev Cell* 5:351–8.

38. Huang, J., Deng, K., Wu, H., Liu, Z., Chen, Z., Cao, S., Zhou, L., Ye, X., Keefe, D. L., and Liu, L. 2008. Efficient production of mice from embryonic stem cells injected into four- or eight-cell embryos by piezo micromanipulation. *Stem Cells* 26:1883–90.

39. Hwang, W. S., Roh, S. I., Lee, B. C., Kang, S. K., Kwon, D. K., Kim, S., Kim, S. J., Park, S. W., Kwon, H. S., Lee, C. K., Lee, J. B., Kim, J. M., Ahn, C., Paek, S. H., Chang, S. S., Koo, J. J., Yoon, H. S., Hwang, J. H., Hwang, Y. Y., Park, Y. S., Oh, S. K., Kim, H. S., Park, J. H., Moon, S. Y., and Schatten, G. 2005. Patient-specific embryonic stem cells derived from human SCNT blastocysts. *Science* 308:1777–83.

40. Hwang, W. S., Ryu, Y. J., Park, J. H., Park, E. S., Lee, E. G., Koo, J. M., Jeon, H. Y., Lee, B. C., Kang, S. K., Kim, S. J., Ahn, C., Hwang, J. H., Park, K. Y., Cibelli, J. B., and Moon, S. Y. 2004. Evidence of a pluripotent human embryonic stem cell line derived from a cloned blastocyst. *Science* 303:1669–74.

41. Jiang, H., Sun, B., Wang, W., Zhang, Z., Gao, F., Shi, G., Cui, B., Kong, X., He, Z., Ding, X., Kuang, Y., Fei, J., Sun, Y. J., Feng, Y., and Jin, Y. 2007. Activation of paternally expressed imprinted genes in newly derived germline-competent mouse parthenogenetic embryonic stem cell lines. *Cell Res* 17:792–803.

42. Kang, L., Wang, J., Zhang, Y., Kou, Z., and Gao, S. 2009. iPS cells can support full-term development of tetraploid blastocyst-complemented embryos. *Cell Stem Cell* 5:135–8.

43. Kaufman, M. H., Barton, S. C., and Surani, M. A. 1977. Normal postimplantation development of mouse parthenogenetic embryos to the forelimb bud stage. *Nature* 265:53–5.

44. Kim, K., Lerou, P., Yabuuchi, A., Lengerke, C., Ng, K., West, J., Kirby, A., Daly, M. J., and Daley, G. Q. 2007a.

Histocompatible embryonic stem cells by parthenogenesis. *Science* 315:482–6.

45. Kim, K., Ng, K., Rugg-Gunn, P. J., Shieh, J. H., Kirak, O., Jaenisch, R., Wakayama, T., Moore, M. A., Pedersen, R. A., and Daley, G. Q. 2007b. Recombination signatures distinguish embryonic stem cells derived by parthenogenesis and somatic cell nuclear transfer. *Cell Stem Cell* 1:346–52.

46. Klimanskaya, I., Chung, Y., Becker, S., Lu, S. J., and Lanza, R. 2006. Human embryonic stem cell lines derived from single blastomeres. *Nature* 444, 481–5.

47. Kline, D., and Kline, J. T. 1992. Repetitive calcium transients and the role of calcium in exocytosis and cell cycle activation in the mouse egg. *Dev Biol* 149:80–9.

48. Lampton, P. W., Crooker, R. J., Newmark, J. A., and Warner, C. M. 2008. Expression of major histocompatibility complex class I proteins and their antigen processing chaperones in mouse embryonic stem cells from fertilized and parthenogenetic embryos. *Tissue Antigens* 72:448–57.

49. Lengerke, C., Kim, K., Lerou, P., and Daley, G. Q. 2007. Differentiation potential of histocompatible parthenogenetic embryonic stem cells. *Ann N Y Acad Sci* 1106:209–18.

50. Lerou, P. H., and Daley, G. Q. 2005. Therapeutic potential of embryonic stem cells. *Blood Rev* 19:321–31.

51. Lerou, P. H., Yabuuchi, A., Huo, H., Miller, J. D., Boyer, L. F., Schlaeger, T. M., and Daley, G. Q. 2008a. Derivation and maintenance of human embryonic stem cells from poor-quality in vitro fertilization embryos. *Nat Protoc* 3:923–33.

52. Lerou, P. H., Yabuuchi, A., Huo, H., Takeuchi, A., Shea, J., Cimini, T., Ince, T. A., Ginsburg, E., Racowsky, C., and Daley,

G. Q. 2008b. Human embryonic stem cell derivation from poor–quality embryos. *Nat Biotechnol* 26:212–4.

53. Levy, D. E., and Lee, C. K. 2002. What does *STAT3* do? *J Clin Invest* 109:1143–8.

54. Li, C., Chen, Z., Liu, Z., Huang, J., Zhang, W., Zhou, L., Keefe, D. L., and Liu, L. 2009a. Correlation of expression and methylation of imprinted genes with pluripotency of parthenogenetic embryonic stem cells. *Hum Mol Genet* 18:2177–87.

55. Li, J., Liu, X., Wang, H., Zhang, S., Liu, F., Wang, X., and Wang, Y. 2009b. Human embryos derived by somatic cell nuclear transfer using an alternative enucleation approach. *Cloning Stem Cells.* 11:39–50.

56. Liang, J., Wan, M., Zhang, Y., Gu, P., Xin, H., Jung, S. Y., Qin, J., Wong, J., Cooney, A. J., Liu, D., and Songyang, Z. 2008. Nanog and *Oct4* associate with unique transcriptional repression complexes in embryonic stem cells. *Nat Cell Biol.* 10:731–739

57. Lin, G., OuYang, Q., Zhou, X., Gu, Y., Yuan, D., Li, W., Liu, G., Liu, T., and Lu, G. 2007. A highly homozygous and parthenogenetic human embryonic stem cell line derived from a one-pronuclear oocyte following in vitro fertilization procedure. *Cell Res* 17:999–1007.

58. Liu, L., and Keefe, D. L. 2007. Nuclear transfer methods to study aging. *Methods Mol Biol* 371:191–207.

59. Liu, L., Trimarchi, J. R., and Keefe, D. L. 2002. Haploidy but not parthenogenetic activation leads to increased incidence of apoptosis in mouse embryos. *Biol Reprod* 66:204–10.

60. Mai, Q., Yu, Y., Li, T., Wang, L., Chen, M. J., Huang, S. Z., Zhou, C., and Zhou, Q. 2007. Derivation of human embryonic stem cell lines from parthenogenetic blastocysts. *Cell Res* 17:1008–19.

61. Mann, J. R., Gadi, I., Harbison, M. L., Abbondanzo, S. J., and Stewart, C. L. 1990. Androgenetic mouse embryonic stem cells are pluripotent and cause skeletal defects in chimeras: implications for genetic imprinting. *Cell* 62:251–60.

62. Mann, J. R., and Stewart, C. L. 1991. Development to term of mouse androgenetic aggregation chimeras. *Development* 113:1325–33.

63. Marson, A., Levine, S. S., Cole, M. F., Frampton, G. M., Brambrink, T., Johnstone, S., Guenther, M. G., Johnston, W. K., Wernig, M., Newman, J., Calabrese, J. M., Dennis, L. M., Volkert, T. L., Gupta, S., Love, J., Hannett, N., Sharp, P. A., Bartel, D. P., Jaenisch, R., and Young, R. A. 2008. Connecting microRNA genes to the core transcriptional regulatory circuitry of embryonic stem cells. *Cell* 134:521–33.

64. Martin, G. R. 1981. Isolation of a pluripotent cell line from early mouse embryos cultured in medium conditioned by teratocarcinoma stem cells. *Proc Natl Acad Sci USA* 78:7634–8.

65. Masui, S., Nakatake, Y., Toyooka, Y., Shimosato, D., Yagi, R., Takahashi, K., Okochi, H., Okuda, A., Matoba, R., Sharov, A. A., Ko, M. S., and Niwa, H. 2007. Pluripotency governed by Sox2 via regulation of Oct3/4 expression in mouse embryonic stem cells. *Nat Cell Biol* 9:625–35.

66. Meissner, A., and Jaenisch, R. 2006. Generation of nuclear transfer-derived pluripotent ES cells from cloned Cdx2-deficient blastocysts. *Nature* 439:212–5.

67. Melton, C., Judson, R. L., and Blelloch, R. 2010. Opposing microRNA families regulate self-renewal in mouse embryonic stem cells. *Nature* 463:621–6.

68. Miura, K., Okada, Y., Aoi, T., Okada, A., Takahashi, K., Okita, K., Nakagawa, M., Koyanagi, M., Tanabe, K., Ohnuki, M., Ogawa, D., Ikeda, E., Okano, H., and Yamanaka, S. 2009. Variation in the safety of induced pluripotent stem cell lines. *Nat Biotechnol.* 27:743–745

69. Niwa, H., Burdon, T., Chambers, I., and Smith, A. 1998. Self-renewal of pluripotent embryonic stem cells is mediated via activation of *STAT3*. *Genes Dev* 12:2048–60.

70. Niwa, H., Miyazaki, J., and Smith, A. G. 2000. Quantitative expression of Oct-3/4 defines differentiation, dedifferentiation or self-renewal of ES cells. *Nat Genet* 24:372–6.

71. Niwa, H., Ogawa, K., Shimosato, D., and Adachi, K. 2009. A parallel circuit of LIF signalling pathways maintains pluripotency of mouse ES cells. *Nature* 460:118–22.

72. Noggle, S., Fung, H. L., Gore, A., Martinez, H., Satriani, K. C., Prosser, R., Oum, K., Paull, D., Druckenmiller, S., Freeby, M., Greenberg, E., Zhang, K., Goland, R., Sauer, M. V., Leibel, R. L., and Egli, D. 2011. Human oocytes reprogram somatic cells to a pluripotent state. *Nature* 478:70–5.

73. Ogawa, H., Shindo, N., Kumagai, T., Usami, Y., Shikanai, M., Jonwn, K., Fukuda, A., Kawahara, M., Sotomaru, Y., Tanaka, S., Arima, T., and Kono, T. 2009. Developmental ability of trophoblast stem cells in uniparental mouse embryos. *Placenta* 30:448–56.

74. Pan, G., and Thomson, J. A. 2007. Nanog and transcriptional networks in embryonic stem cell pluripotency. *Cell Res* 17:42–9.

75. Pesce, M., and Scholer, H. R. 2000. Oct-4: control of toti-potency and germline determination. *Mol Reprod Dev* 55:452–7.

76. Qi, X., Li, T. G., Hao, J., Hu, J., Wang, J., Simmons, H., Miura, S., Mishina, Y., and Zhao, G. Q. 2004. BMP4 supports self-renewal of embryonic stem cells by inhibiting mitogen-activated protein kinase pathways. *Proc Natl Acad Sci USA* 101:6027–32.

77. Revazova, E. S., Turovets, N. A., Kochetkova, O. D., Agapo-va, L. S., Sebastian, J. L., Pryzhkova, M. V., Smolnikova, V. I., Kuzmichev, L. N., and Janus, J. D. 2008. HLA homozygous stem cell lines derived from human parthenogenetic blas-tocysts. *Cloning Stem Cells* 10:1–24.

78. Revazova, E. S., Turovets, N. A., Kochetkova, O. D., Kinda-rova, L. B., Kuzmichev, L. N., Janus, J. D., and Pryzhkova, M. V. 2007. Patient-specific stem cell lines derived from human parthenogenetic blastocysts. *Cloning Stem Cells* 9:432–49.

79. Ringrose, L., and Paro, R. 2004. Epigenetic regulation of cel-lular memory by the Polycomb and Trithorax group proteins. *Annu Rev Genet* 38:413–43.

80. Sato, N., Meijer, L., Skaltsounis, L., Greengard, P., and Brivan-lou, A. H. 2004. Maintenance of pluripotency in human and mouse embryonic stem cells through activation of WnT sig-naling by a pharmacological GSK-3-specific inhibitor. *Nat Med* 10:55–63.

81. Shao, H., Wei, Z., Wang, L., Wen, L., Duan, B., Mang, L., and Bou, S. 2007. Generation and characterization of mouse parthenogenetic embryonic stem cells containing ge-nomes from non-growing and fully grown oocytes. *Cell Biol Int* 31:1336–44.

82. Sritanaudomchai, H., Pavasuthipaisit, K., Kitiyanant, Y., Kupradinun, P., Mitalipov, S., and Kusamran, T. 2007. Characterization and multilineage differentiation of embryonic stem cells derived from a buffalo parthenogenetic embryo. *Mol Reprod Dev* 74:1295–302.

83. Stojkovic, M., Stojkovic, P., Leary, C., Hall, V. J., Armstrong, L., Herbert, M., Nesbitt, M., Lako, M., and Murdoch, A. 2005. Derivation of a human blastocyst after heterologous nuclear transfer to donated oocytes. *Reprod Biomed Online* 11:226–31.

84. Strumpf, D., Mao, C. A., Yamanaka, Y., Ralston, A., Chawengsaksophak, K., Beck, F., and Rossant, J. 2005. Cdx2 is required for correct cell fate specification and differentiation of trophectoderm in the mouse blastocyst. *Development* 132:2093–102.

85. Suzuki, A., Raya, A., Kawakami, Y., Morita, M., Matsui, T., Nakashima, K., Gage, F. H., Rodriguez-Esteban, C., and Izpisua Belmonte, J. C. 2006. Nanog binds to Smad1 and blocks bone morphogenetic protein-induced differentiation of embryonic stem cells. *Proc Natl Acad Sci USA* 103:10294–9.

86. Swann, K., and Ozil, J. P. 1994. Dynamics of the calcium signal that triggers mammalian egg activation. *Int Rev Cytol* 152:183–222.

87. Tada, M., Takahama, Y., Abe, K., Nakatsuji, N., and Tada, T. 2001. Nuclear reprogramming of somatic cells by in vitro hybridization with ES cells. *Curr Biol* 11:1553–8.

88. Takahashi, K., Tanabe, K., Ohnuki, M., Narita, M., Ichisaka, T., Tomoda, K., and Yamanaka, S. 2007. Induction of pluripotent stem cells from adult human fibroblasts by defined factors. *Cell* 131:861–72.

89. Takahashi, K., and Yamanaka, S. 2006. Induction of pluripotent stem cells from mouse embryonic and adult fibroblast cultures by defined factors. *Cell* 126:663–76.

90. Tam, W. L., Lim, C. Y., Han, J., Zhang, J., Ang, Y. S., Ng, H. H., Yang, H., and Lim, B. 2008. T-cell factor 3 regulates embryonic stem cell pluripotency and self-renewal by the transcriptional control of multiple lineage pathways. *Stem Cells.* 26:2019–31.

91. Tanaka, M., Hadjantonakis, A. K., Vintersten, K., and Nagy, A. 2009. Aggregation chimeras: combining ES cells, diploid, and tetraploid embryos. *Methods Mol Biol* 530:287–309.

92. Teramura, T., Onodera, Y., Murakami, H., Ito, S., Mihara, T., Takehara, T., Kato, H., Mitani, T., Anzai, M., Matsumoto, K., Saeki, K., Fukuda, K., Sagawa, N., and Osoi, Y. 2009. Mouse androgenetic embryonic stem cells differentiated to multiple cell lineages in three embryonic germ layers in vitro. *J Reprod Dev* 55:283–92.

93. Thomson, J. A., Itskovitz-Eldor, J., Shapiro, S. S., Waknitz, M. A., Swiergiel, J. J., Marshall, V. S., and Jones, J. M. 1998. Embryonic stem cell lines derived from human blastocysts. *Science* 282:1145–7.

94. Thomson, J. A., Kalishman, J., Golos, T. G., Durning, M., Harris, C. P., Becker, R. A., and Hearn, J. P. 1995. Isolation of a primate embryonic stem cell line. *Proc Natl Acad Sci USA* 92:7844–8.

95. Toth, S., Huneau, D., Banrezes, B., and Ozil, J. P. 2006. Egg activation is the result of calcium signal summation in the mouse. *Reproduction* 131:27–34.

96. Tseng, A. S., Engel, F. B., and Keating, M. T. 2006. The GSK-3 inhibitor BIO promotes proliferation in mammalian cardiomyocytes. *Chem Biol* 13:957–63.

97. Tuch, B. E. 2006. Stem cells—a clinical update. *Aust Fam Physician* 35:719–21.

98. Vogel, G. 2008. Breakthrough of the year. Reprogramming Cells. *Science* 322:1766–7.

99. von Bubnoff, A., and Cho, K. W. 2001. Intracellular BMP signaling regulation in vertebrates: pathway or network? *Dev Biol* 239:1–14.

100. Vrana, K. E., Hipp, J. D., Goss, A. M., McCool, B. A., Riddle, D. R., Walker, S. J., Wettstein, P. J., Studer, L. P., Tabar, V., Cunniff, K., Chapman, K., Vilner, L., West, M. D., Grant, K. A., and Cibelli, J. B. 2003. Nonhuman primate parthenogenetic stem cells. *Proc Natl Acad Sci USA* 100 Suppl 1:11911–6.

101. Waite, L., and Nindl, G. 2003. Human embryonic stem cell research: an ethical controversy in the US & Germany. *Biomed Sci Instrum* 39:567–72.

102. Wakayama, T., Perry, A. C., Zuccotti, M., Johnson, K. R., and Yanagimachi, R. 1998. Full-term development of mice from enucleated oocytes injected with cumulus cell nuclei. *Nature* 394:369–74.

103. Wang, S., Tang, X., Niu, Y., Chen, H., Li, B., Li, T., Zhang, X., Hu, Z., Zhou, Q., and Ji, W. 2007. Generation and characterization of rabbit embryonic stem cells. *Stem Cells* 25:481–9.

104. Wilmut, I., Schnieke, A. E., McWhir, J., Kind, A. J., and Campbell, K. H. 1997. Viable offspring derived from fetal and adult mammalian cells. *Nature* 385:810–3.

105. Wobus, A. M., and Boheler, K. R. 2005. Embryonic stem cells: prospects for developmental biology and cell therapy. *Physiol Rev* 85:635–78.

106. Wray, J., Kalkan, T., Gomez-Lopez, S., Eckardt, D., Cook, A., Kemler, R., and Smith, A. 2011. Inhibition of glycogen synthase kinase-3 alleviates Tcf3 repression of the pluripotency network and increases embryonic stem cell resistance to differentiation. *Nat Cell Biol* 13:838–45.

107. Yi, F., Pereira, L., and Merrill, B. J. 2008. Tcf3 functions as a steady state limiter of transcriptional programs of mouse embryonic stem cell self renewal. *Stem Cells*. 26:1951–60.

108. Yi, F., Pereira, L., Hoffman, J. A., Shy, B. R., Yuen, C. M., Liu, D. R., and Merrill, B. J. 2011. Opposing effects of Tcf3 and Tcf1 control WnT stimulation of embryonic stem cell self-renewal. *Nat Cell Biol* 13:762–70.

109. Ying, Q. L., Nichols, J., Chambers, I., and Smith, A. 2003. BMP induction of Id proteins suppresses differentiation and sustains embryonic stem cell self-renewal in collaboration with *STAT3. Cell* 115:281–92.

110. Ying, Q. L., Wray, J., Nichols, J., Batlle-Morera, L., Doble, B., Woodgett, J., Cohen, P., and Smith, A. 2008. The ground state of embryonic stem cell self-renewal. *Nature* 453:519–23.

111. Yu, J., Vodyanik, M. A., Smuga-Otto, K., Antosiewicz-Bourget, J., Frane, J. L., Tian, S., Nie, J., Jonsdottir, G. A., Ruotti, V., Stewart, R., Slukvin, II, and Thomson, J. A. 2007. Induced pluripotent stem cell lines derived from human somatic cells. *Science* 318:1917–20.

112. Zhang, X., Stojkovic, P., Przyborski, S., Cooke, M., Armstrong, L., Lako, M., and Stojkovic, M. 2006. Derivation of human embryonic stem cells from developing and arrested embryos. *Stem Cells* 24:2669–76.

113. Zhao, X. Y., Li, W., Lv, Z., Liu, L., Tong, M., Hai, T., Hao, J., Guo, C. L., Ma, Q. W., Wang, L., Zeng, F., and Zhou, Q. 2009. iPS cells produce viable mice through tetraploid complementation. *Nature*. 416:86–90

Chapter 3

Isolation and Maintenance of Adult Cardiac Stem Cells

Michael Bauer [1], Ahmad F. Bayomy [1,2], and Ronglih Liao [1]

[1] Cardiac Muscle Research Laboratory, Cardiovascular Division, Department of Medicine, Brigham and Women's Hospital, Harvard Medical School, Boston, MA
[2] University of Washington School of Medicine, Seattle, WA

For correspondence:
Ronglih Liao, PhD
Cardiac Muscle Research Laboratory
Cardiovascular Division
Department of Medicine
Brigham and Women's Hospital
Harvard Medical School
77 Avenue Louis Pasteur, NRB 431
Boston, MA 02115

E-mail: rliao@rics.bwh.harvard.edu
Phone: (617) 525-4854
Fax: (617) 525-4868

1. Introduction

Stem cells are known for their remarkable capacity to renew indefinitely and for their ability to differentiate, upon stimulation, into the body's many cell types. Given the recent advances in knowledge surrounding stem cells and their potential for cellular plasticity, regenerative medicine has evolved as a viable and exciting therapeutic strategy for delaying, stopping, or—most impressively—reversing and thus curing disease. As such, stem cell research has emerged as one of the most fascinating areas of modern biology. However, much work remains toward achieving the ultimate goal of complete tissue regeneration.

Stem cells can generally be grouped into two categories based on their tissues of origin: embryonic stem cells and non-embyronic stem cells. Embryonic stem cells are derived from the early inner cell mass of the embryo, or blastocyst. Non-embryonic stem cells are derived from somatic (non-germline) or adult tissues and organs.

In this chapter, we will focus on adult stem cells—with an emphasis on cells from cardiac tissue. We aim to provide you with a general overview of the current state of adult cardiac stem cell biology, including a discussion of the definition, isolation, and maintenance of stem cells. In addition, we intend to highlight the potential for use in cardiovascular regeneration while recognizing existing limitations.

2. What Are Adult Stem Cells?

Adult and somatic stem cells are primitive, undifferentiated cells residing in adult organs and are surrounded by organ-specific, differentiated cells. Such stem cells possess the capability for self-renewal and differentiation into some, if not all, of the given cell types within a respective organ or tissue. The primary role of adult stem cells, therefore, is thought to be the maintenance of cellular homeostasis during the normal aging process or in response to cellular stress or injury.

Research on adult stem cells dates back to the early 1950s when it was first discovered that bone marrow contains populations of primitive stem cells.[1,2] These hematopoietic stem cells, known to form all blood cell types in the body, have been used extensively in bone marrow transplantation and have greatly impacted medical practice over the last several decades.[3] Following the initial discovery of bone marrow–derived stem cells, scientists have found adult stem cells throughout the body—including in those organs and tissues once believed to be terminally differentiated (e.g., the brain and heart).[4]

The identification of adult stem cells has led scientists and clinicians to question whether it is possible to harness these cells for the treatment of disease. Potentially, such a therapeutic approach could surpass the impact seen with bone marrow transplantation, since allogeneic (person-to-person) transplants carry a risk of adverse immune reactions. Therefore, if possible, it would be advantageous to transplant a person's own stem cells for treatment of a specific disease.

Although the developmental origin of most adult stem cells is not fully known, the therapeutic potential remains significant. Scientists and researchers worldwide have raced to better understand the regulation of adult stem cells in all organs and tissues, and the collective hope is to identify the ideal adult stem cells for different therapies.

3. Can the Adult Heart Regenerate Itself?

Until recently, the heart was described as a terminally differentiated organ with no capacity for self-regeneration. The long-lasting dogma held that the number of cardiac muscle cells was determined shortly after birth, and any injury resulting in the loss of cardiac muscle cells was an irreversible process. Emerging data in both animal models and human tissue have demonstrated that the adult heart may possess a noticeable degree of regenerative capacity.[5,6] Still, the source(s) of new cardiac muscle cells remains a focus of study. It has been proposed that newly-generated cardiac muscle cells arise through one or more mechanisms, including but not limited to: (1) the division of terminally differentiated cardiac muscle cells which re-enter the cell cycle, (2) de novo differentiation from resident cardiac stem or progenitor cells, and (3) trans-differentiation from stem cell pools outside the heart. Researchers across the globe are investigating these and other potential mechanisms for cardiac cell renewal, and a definitive answer is expected in the near future.

4. What Are Adult Cardiac Stem Cells?

Within the field of stem cell biology, cardiac stem cells have taken center stage and created a buzz surrounding cell-based therapy for the treatment of heart disease. Not only is the heart a vital organ, but heart disease is also the leading cause of death in industrialized countries.[7] Over the past decade, researchers have begun to shed light on the notion of cardiac regeneration through the discovery of several populations of cardiac stem or progenitor cells from the adult heart. For the sake of discussion, we will describe a cell's so-called stem-like characteristics through the example of cardiac stem cells.

87

To be qualified as a bona fide group of adult cardiac stem cells, a cell population should fulfill three criteria. First, cells must be isolated from adult cardiac tissue and be free of contamination from circulating blood cells or infiltrative cells. Second, the isolated cells must be expanded clonally. Third, the cells must be capable of self-renewal and differentiation into not only the major cell types of the heart—cardiac muscle cells, endothelial cells, and smooth muscle cells—but also other cell lineages.

Thus far, different cardiac stem cell populations share in common that they have been isolated from postnatal heart tissue and possess the ability to differentiate into the three major cardiac cell lineages. It is important to note that only some populations of cardiac stem cells are demonstrated to differentiate into multiple cell types from a single clone. As such, it is possible that various differentiated cells may instead be derived from different sub-populations of progenitor cells within the larger population. In other words, cardiac muscle cells may be derived from dedicated cardiac progenitor cells and endothelial cells from dedicated endothelial progenitor cells, rather than a single stem cell producing one or both cell types. We use the term "cardiac stem/progenitor cells" to describe the various populations of cells that have been shown to differentiate into all three major cell types within the heart.

The different populations of adult cardiac stem/progenitor cells can generally be distinguished by the methods employed for their isolation: expression of specific surface markers, functional phenotype, or unique properties in culture. Despite the different methodologies for isolation, there is some degree of overlap among all, namely in the expression of common cell surface markers. Even though the nature of the relationships between these populations of cardiac stem/progenitor cells is not yet fully understood, the fact remains that all carry potential for therapeutic cardiac regeneration. The major characterization of these reported cardiac stem/progenitor cells has been extensively reviewed;[8] herein, we center our attention on the methods used to isolate and cultivate these cells for use in laboratory research and therapeutic applications.

5. How Are Adult Cardiac Stem Cells Isolated?

As summarized in table 1, stem/progenitor cells can be isolated from adult cardiac tissue based on the expression of certain surface markers (*c-Kit* or *Sca-1*) or functional phenotype (Hoechst dye 33342 efflux) intrinsic to these cells. Alternatively, cardiac stem/progenitor cells can also be isolated by their unique ability to form spheres in culture (cardiospheres). Accordingly, we base our discussion of populations of cardiac stem/progenitor cells on the respective isolation methods. It is important to mention that other cardiac progenitor cell types have been identified, including isl-1+,[9] wt-1+,[10] and tbx-18+[11] cells. These particular cell types are important in neonatal development and cannot be found in the adult heart, so we will not discuss them here.

5.1 Isolation using surface markers

To date, two cardiac stem/progenitor cell populations have been identified and isolated by the expression of specific proteins on the cell surface: *Sca-1* positive cells and *c-Kit* positive cells.[12,13] *Sca-1* (*stem cell antigen-1*) is a common cell surface protein that was initially found to be highly expressed in hematopoietic stem cells. Subsequently, *Sca-1* expression has been identified in stem/progenitor cells from an array of adult tissues, including mammary glands,[14] skeletal muscle,[15] testis,[16] and cardiac tissue.[13] *Sca-1* positive cardiac stem/progenitor cells were first identified and characterized by Schneider and colleagues using the heart of the adult mouse.[13] Similarly, *c-Kit* (also known as *CD117*) functions as a cell surface receptor for stem cell factor. *c-Kit* is a common stem cell marker for a variety of adult stem cells, including hematopoietic and cardiac stem cells. Anversa and colleagues[12] were the first to isolate and thoroughly characterize the *c-Kit* population of cells from cardiac tissue of humans and other species. Currently, *c-Kit* positive cells are the only population of adult cardiac stem/progenitor cells qualified as true stem cells as they have been shown to satisfy the three criteria of self-renewal, clonal expansion, and multipotency.[12]

Table 1. Overview of selected stem/progenitor cell types by isolation technique.

Cell Type	Isolation Technique	Species	Digestion Enzyme(s)
Ckit+	Surface Marker[12]	Human Mouse Dog Rat Pig	Collagenase[21]
Sca-1+	Surface Marker[13]	Mouse Dog	Collagenase[13]
CSP	Phenotype (Hoechst efflux)[22,23]	Human Mouse Pig	Collagenase and dispase[23] Pronase[24]
Cardio-sphere	Phenotype (Outgrowth)[19]	Human Dog Mouse	Collagenase and trypsin[19]

Following mechanical and enzymatic dissociation, cell debris and larger pieces of tissue can be removed by filtering the tissue/cell suspension through several meshes with gradually

decreasing pore size. Since the markers of interest are proteins expressed on the surfaces of cells, antibodies specific to
these proteins can readily be used to select for the desired
cell population. The antibodies can either be conjugated to
magnetic beads for magnetic cell sorting or to fluorophores
for fluorescence-activated cell sorting (FACS).

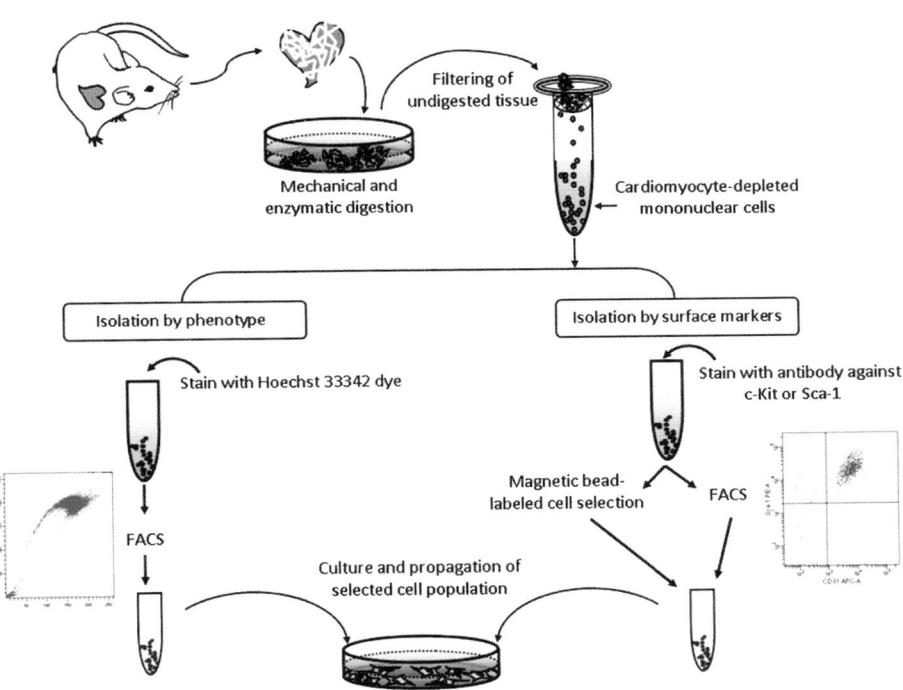

**Figure 1 illustrates a general schematic for the isolation of tissue stem/
progenitor cells. The first step, common to all isolation techniques, is to
extract the resident stem/progenitor cells from the surrounding tissue.
This usually requires mechanical and enzymatic dissociation in order to
release the cells from the extracellular matrix (ECM) network of cardiac
tissue. While each laboratory may utilize a slightly different cocktail of
enzymes, typical enzymes capable of digesting ECM include collagenase, trypsin, and dispase (table 1). In high concentrations, these enzymes also serve to eliminate mature cardiac muscle cells from the cell
suspension.**

For the approach using conjugated magnetic beads, the cell suspension is passed through a column inside of a magnet. If antibodies are bound to a cell, the cell is retained inside the magnetic field. After the entire suspension is passed through the column, the magnet is removed and the retained cells are acquired through washing.

In the FACS approach, a thin stream of cells is passed through a laser beam that excites the fluorophores and induces them to emit light at a specific wavelength. This signal is recorded by detectors and then processed by a computer program that sorts cells based on fluorescence and size. While magnetic bead–labeled cell selection can select for cells expressing individual proteins of interest, FACS has the unique ability to select cells based on multiple markers (i.e., Sca-1+ CD45-). Thus, FACS may be used to exclude cells bearing certain lineage markers.

5.2 Isolation by phenotype

Cardiac stem/progenitor cells have the unique ability to efflux certain DNA-binding dyes, and this function is utilized to isolate and define side population (SP) cardiac stem/progenitor cells. This method was first utilized by Goodell and colleagues to identify enriched hematopoietic stem cells populations in bone marrow.[17] Capitalizing on the dye-effluxing property, scientists have gone on to identify stem/progenitor cells from many organs and tissues including the heart.[18]

The ability of cells to efflux a DNA-binding dye, Hoechst 33342, is conferred by specific transporters on the cell surface, such as *Bcrp1* (*breast cancer resistant protein 1*) or *Mdr1* (*multi drug resistant protein 1*). The transporters are express on many primitive cell types, and it is believed that their original purpose was to remove intracellular toxins or other potentially harmful substances.

As shown in figure 1, identification of SP cells relies solely on FACS. First, the total cell number is determined via a cell-counting device, or hemacytometer. Recall that the mononuclear cells were depleted of cardiac muscle cells through enzymatic actions and filtering. The cells are then stained with Hoechst 33342 and subjected to FACS to distinguish, or gate, those cells effluxing the dye. The ratio of the concentration of Hoechst dye to the number of cells is critically important to capture the true SP population. If the dye concentration is high relative to the cell number, the cells are overwhelmed, and the overall efflux ability is diminished. Conversely, if the ratio of dye concentration to cells is too low, then the percentage of SP cells will be artificially high. Verapamil, a calcium channel-blocking drug which inhibits cell surface transporters, is commonly used to aid in gating the correct population of SP cells from the main cell population. After Hoechst staining, antibodies can be used to include or exclude cells based on the expression of additional surface proteins.

5.3 Isolation by cell culture

Cardiospheres are characterized by the unique appearance of spheres of cells when cultured under pre-defined conditions.[19] The mechanical and enzymatic dissociation used for the isolation of cardiospheres is slightly different from that used to isolate *Sca-1*, *c-Kit*, or SP cells. Typically, heart tissue is diced into pieces with a volume of roughly 1–2 cubic millimeters; these pieces are immediately digested with trypsin and collagenase and cultured on a fibronectin-coated plate. After several days in culture, fibroblast cells grow from the pieces of tissue. With continued culture for several more days, phase bright cells appear on top of the fibroblast layer (named as such due to their bright appearance on phase contrast microscopy). The tissue pieces are removed, and the phase bright cells are carefully harvested and plated into culture dishes in a specific medium. The cells form characteristic spheres in this environment, with each sphere thought to comprise clones of a single cell.

Note that cardiospheres commonly express stem cell markers, such as c-Kit and Sca-1. Compared to other populations of adult cardiac stem/progenitor cells, cardiospheres are the most heterogeneous. It has yet to be determined whether the cell heterogeneity is beneficial or deleterious with regard to use in cell-based therapy.

6. How Are Adult Cardiac Stem/ Progenitor Cells Maintained and Multiplied In Vitro?

Adult cardiac stem/progenitor cell numbers are found in relatively low quantities in vivo. Thus, isolated cells need to be cultivated and multiplied, or expanded, to obtain sufficient numbers for experimental or therapeutic use. Cell expansion occurs over several iterations, termed passages, and is achieved by culturing cells in optimized conditions. Most importantly, this involves a unique medium for each of the identified adult stem/progenitor cell populations (table 2). Otherwise, the cultivation of adult stem/progenitor cells in vitro requires the same standards as any typical cell culture procedure. It is relevant to note that adult stem/progenitor cells, unlike embryonic stem cells, do not require a feeder layer on which to grow. Feeder layers consist of inactivated cells used to support the proliferation of stem cells, and they may pose a risk of infection if derived from exogenous sources (e.g., animals or other humans).[20]

Table 2. Culture conditions for selected stem/progenitor cell types.

Cell Type	Base Medium	Serum	Additional conditions
Ckit+	F12K medium	5% FBS	bFGF and LIF [12]
Sca-1+	Medium-199	10% FBS	Fibronectin-coated dish [13]
CSP	alphaMEM	20% FBS	L-glutamine [23]
	Iscove's MDM	10% FBS	Gelatin-coated dish [25]
		2% FBS	None [24]
Cardio-sphere	35% IMDM 65% DMEM-Ham F-12 2% B27	10% FBS	2-mercaptoethanol, EGF, bFGF, cardiotrophin-1, trombin, l-glutamine [19]

Culture conditions, particularly the composition of the medium, strongly influence both the quantity and the quality of the cultured stem/progenitor cells. The primary aim is to expand the cells and simultaneously maintain their intrinsic capacity for multipotent differentiation. To maintain cells in their primitive and undifferentiated state, culture media should contain serum, growth factors, glucose, and standard essential amino acids. Maintaining the physiologic pH of the culture media is important, and this can be monitored via a color indicator, phenol red, found in most media.

For culturing attached cells, the degree of cellular confluency is important and can be estimated based on the amount of space covered by cells relative to the unoccupied space. Cells should be passaged when they reach a threshold level of confluency order to prevent over-crowding and cell death. Similarly, cardiospheres are only able to grow to a certain size, so they are frequently dissolved using trypsin and re-cultured for expansion of cell numbers.

Signs of unhealthy culture conditions include rounded cells, floating cells, and giant egg-shaped cells, the latter being a sign of cells entering senescence. Cells should be examined daily through visual inspection as well as at each passage through quantitative measures of stem capability and overall cell well-being: telomere length, specific gene and protein expression, and/or functional assays.

It is appreciated that culture conditions cannot fully replicate the in situ cellular environment, particularly when cells are situated on a two-dimensional petri dish surface. Consequently, and with advances in tissue bioengineering, scientists are working to create culture conditions in three dimensions that better mimic the tissue environment.

7. Closing Remarks

We have only begun to witness the tip of the iceberg with regard to the enormous impact that stem cell research can have on regenerative medicine. The discovery of resident adult cardiac stem/progenitor cells provides us with the tangible hope for complete cardiac regeneration. However, the pursuit of this therapeutic goal is limited by our incomplete understanding of the mechanisms that regulate the fate of cardiac stem/progenitor cells in vivo.

Acknowledgements

We would like to thank Melissa Teng for her artistic contributions.

References

1. Lorenz E., Congdon C., and Uphoff D. 1952. Modification of acute irradiation injury in mice and guinea-pigs by bone marrow injections. *Radiology.* 58(6):863–77.

2. Main J. M. and Prehn R. T. 1955. Successful skin homografts after the administration of high dosage X radiation and homologous bone marrow. *Journal of the National Cancer Institute* 15(4):1023–9.

3. Thomas E. D. 2005. Bone marrow transplantation from the personal viewpoint. *International Journal of Hematology* 81(2):89–93.

4. Alison M. R. and Islam S. 2009. Attributes of adult stem cells. *The Journal of Pathology* 217(2):144–60.

5. Bergmann O., Bhardwaj R., Bernard S., et al. 2009. Evidence for Cardiomyocyte Renewal in Humans. *Science* 324(5923):98–102.

6. Hsieh P. C., Segers V. F., Davis M. E., et al. 2007. Evidence from a genetic fate-mapping study that stem cells refresh adult mammalian cardiomyocytes after injury. *Nature Medicine* 13(8):970–4.

7.   WHO Statistical Information System, http://www.who.int/
     whosis/en/ (World Health Statistics 2008; accessed September 9, 2009).

8.   Barile L., Messina E., Giacomello A., and Marb E. 2007. Endogenous Cardiac Stem Cells. *Prog Cardiovasc Dis* 50(1):31–48.

9.   Laugwitz K. L., Moretti A., Lam J., et al. 2005. Postnatal isl1+
     cardioblasts enter fully differentiated cardiomyocyte lineages. *Nature* 433(7026):647–53.

10. Zhou B., Ma Q., Rajagopal S., et al. 2008. Epicardial progenitors contribute to the cardiomyocyte lineage in the
     developing heart. *Nature* 454(7200):109–13.

11. Cai C. L., Martin J., Sun Y., et al. 2008. A myocardial lineage
     derives from Tbx18 epicardial cells. *Nature* 454(7200):104–8.

12. Beltrami A. P., Barlucchi L., Torella D., et al. 2003. Adult cardiac stem cells are multipotent and support myocardial regeneration. *Cell* 114(6):763–76.

13. Oh H., Bradfute S. B., Gallardo T. D., et al. 2003. Cardiac
     progenitor cells from adult myocardium: homing, differentiation, and fusion after infarction. *Proceedings of the National Academy of Sciences of the United States of America*
     100(21):12313–8.

14. Welm B. E., Tepera S. B., Venezia T., Graubert T. A., Rosen J.
     M., and Goodell M. A. 2002. Sca-1(pos) cells in the mouse
     mammary gland represent an enriched progenitor cell
     population. *Developmental Biology* 245(1):42–56.

15. Royer C. L., Howell J. C., Morrison P. R., Srour E. F., and Yoder
     M. C. 2002. Muscle-derived CD45-SCA-1+c-kit- progenitor

cells give rise to skeletal muscle myotubes in vitro. In vitro cellular & developmental biology. *Animal* 38(9):512–7.

16. Van Bragt M. P., Ciliberti N., Stanford W. L., De Rooij D. G., and Van Pelt A. M. 2005. LY6A/E (SCA-1) expression in the mouse testis. *Biology of Reproduction* 73(4):634–8.

17. Goodell M. A., Brose K., Paradis G., Conner A. S., and Mulligan R. C. 1996. Isolation and functional properties of murine hematopoietic stem cells that are replicating in vivo. *The Journal of Experimental Medicine* 183(4):1797–806.

18. Challen G. and Little M. 2006. A side order of stem cells: the SP phenotype. *Stem Cells* (Dayton, Ohio) 24(1):3–12.

19. Messina E., De Angelis L., Frati G., et al. 2004. Isolation and expansion of adult cardiac stem cells from human and murine heart. *Circulation Research* 95(9):911–21.

20. Halme D. G. and Kessler D. A. 2006. FDA regulation of stem-cell-based therapies. *The New England journal of Medicine* 355(16):1730–5.

21. Kubo H., Jaleel N., Kumarapeli A., et al. 2008. Increased cardiac myocyte progenitors in failing human hearts. *Circulation* 118(6):649–57.

22. Hierlihy A. M., Seale P., Lobe C. G., Rudnicki M. A., and Megeney L. A. 2002. The post-natal heart contains a myocardial stem cell population. *FEBS letters* 530(1–3):239–43.

23. Pfister O., Mouquet F., Jain M., et al. 2005. CD31- but Not CD31+ cardiac side population cells exhibit functional cardiomyogenic differentiation. *Circulation Research* 97(1):52–61.

24. Martin C. M., Meeson A. P., Robertson S. M., et al. 2004. Persistent expression of the ATP-binding cassette transporter, Abcg2, identifies cardiac SP cells in the developing and adult heart. *Developmental Biology* 265(1):262–75.

25. Oyama T., Nagai T., Wada H., et al. 2007. Cardiac side population cells have a potential to migrate and differentiate into cardiomyocytes in vitro and in vivo. *The Journal of Cell Biology* 176(3):329–41.

Chapter 4

Molecular Mechanisms Underlying Embryonic Stem Cell Self-Renewal

Ming Zhan
Department of Systems Medicine and Bioengineering, The Methodist
Hospital Research Institute, Houston, TX, USA

For correspondence:
The Methodist Hospital Research Institute
6670 Bertner Street
Houston, TX 77030

E-mail: mzhan@tmhs.org
Phone: (713) 441-8939

1. Abstract

Embryonic stem (ES) cells are pluripotent, characterized by the specific ability of self-renewal and differentiation. An understanding of the molecular mechanisms underlying

ES cell pluripotency is essential to realize their potentials in regenerative medicine and science. ES cell pluripotency is regulated by various signaling pathways and biological processes including cell cycle, BMP, TGF-β, WnT, FGF, and Nodal. These pathways are mediated by intrinsic transcription factors such as Oct4, Sox2, and Nanog, which work coordinately to activate genes critical for self-renewal and repress genes associated with cell lineage commitment, determining ES cell identity. Reprogramming somatic cells using these transcription factors can induce ES cell–like pluripotent stem cells. Epigenetic regulation, including histone modifications, DNA methylation, and non-coding RNA regulation, acts as a complementary mechanism to the transcriptional control of ES cell pluripotency. The molecular mechanisms controlling ES cell self-renewal and differentiation are either evolutionary conserved and fundamental, or divergent and species-specific. Herein, we describe major molecular mechanisms governing ES cell pluripotency, with particular emphases on recent discoveries in this field.

2. Introduction

Embryonic stem (ES) cells are pluripotent, with unlimited replication potential and capable of giving rise to all kinds of cell lineages in the embryo and adult. Because of these features, the cells have been extensively used as a biological model for studying development. Additionally, they represent a valuable source for regenerative medicine for treatment of many currently incurable injuries and diseases. ES cells were initially derived from the inner cell mass (ICM) of blastocysts in mice in 1981 [18, 46]. Since then, the same cells have been isolated from other species, including humans and primates [76, 77, 78]. Recently, scientists successfully isolated pluripotent cells through reprogramming adult somatic cells [56, 73, 88]. The resulting induced pluripotent stem (iPS) cells show a high similarity to ES cells in many aspects of molecular and cell biology.

ES cells face two fates in their development: remaining in the pluripotent state by self-renewal or going through differentiation. ES cells move closer to their fate at each cell division by integrating signals from the microenvironment, or so-called stem cell niche, and starting the built-in molecular machineries through a hierarchical structure. The structure is comprised of three major interacting components: (1) signaling pathways receiving and transferring extracellular signals into the cell; (2) the core regulatory network organized by intrinsic transcription factors (TFs) and their co-regulators for integrating and processing intracellular signals; (3) target genes regulated by the core regulatory network [51, 62]. The self-renewal process, which maintains pluripotency without losing the ability to differentiate, is carried out by activating ES cell–specific genes and repressing cell type–specific genes, along with regulation of the cell cycle process. The differentiation process, on the other hand, can be driven by opposite efforts, i.e. activating cell type–specific pathways, repressing self-renewal signals, and altering the cell cycle control. A number of signal pathways and transcription regulators account for self-renewal and differentiation of ES cells. Their complex interactions are the key mechanisms underlying ES cell pluripotency. Understanding the molecular mechanisms is critical for realizing the great potentials of ES cells in medicine and science.

Emerging high-throughput technologies have facilitated large-scale genomics, transcriptomics, proteomics, and epigenetics studies of ES cells. The studies shed light on the understanding of the molecular mechanisms underlying ES cell pluripotency. Progress has particularly been made in identifying key regulatory factors that are critical to ES cell self-renewal and differentiation or essential for generating iPS cells. Cross-species comparison of ES cells further allows

103

identifying conserved and divergent pathways guiding ES cell pluripotency. In this chapter, we present an overview of molecular mechanisms regulating ES cell pluripotency. We first focus on the description of signaling pathways, transcription regulators, and their intricate interplays—crucial for decision-making on ES cell fate. We next discuss regulatory roles of epigenetic modifications during ES cell development. We then present novel discoveries of several core factors that are able to reprogram somatic cells to generate iPS cells. Finally, we briefly introduce evolutionary conservation and divergence of molecular mechanisms controlling ES cell development.

3. Signaling Pathways

3.1 LIF signaling

LIF signaling is critical for self-renewal of mouse ES cells but not essential for human ES cells. LIF is a member of the super-family of cytokines. LIF signaling in mouse ES cells starts with dimerization of LIF receptor (LIFR) and GP130 after the binding of LIF (figure 1). *Janus kinase (JAK)* members are then activated by LIFR and GP130, and in turn phosphorylate *STAT3*. The phosphorylated *STAT3* translocates into the nucleus and promotes the expression of downstream genes. *STAT3* directly regulates the expression of *MYC* and stabilizes a higher level of active *MYC* in ES cells, so as to prevent differentiation [14]. The activated *STAT3* and constitutively expressed *MYC* benefits self-renewal of mouse ES cells, while an opposite status of *STAT3* or *MYC* shifts ES cells to differentiation. However, LIF alone is not sufficient to sustain mouse ES cells at the undifferentiated state in a serum-free culture. In the absence of serum factors, LIF could not prevent differentiation of mouse ES cells, but its combination with bone morphogenetic proteins (BMPs) can keep ES cells at the self-renewal or pluripotency state [87]. The effect of BMP on mouse ES cell self-renewal is LIF-dependent; without LIF, BMP induces differentiation of mouse ES cells.

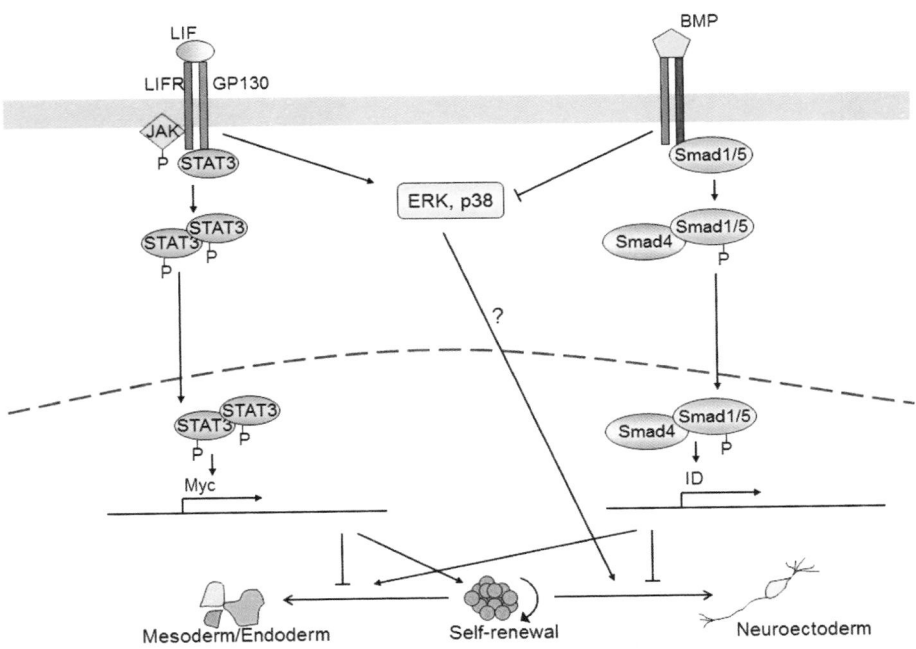

Figure 1. Cooperation between LIF and BMP signaling maintains mouse ES cell pluripotency. LIF signaling blocks the differentiation of mouse ES cells into differentiation through the activation of STAT3, which promotes the transcription of MYC. LIF signaling also activates ERK, which likely promotes the differentiation of ES cells into neural lineages. ID expression induced by BMP can inhibit ES cell differentiation. The activity of ERK can also be suppressed by BMP signaling, relieving negative effects of ERK on pluripotency. Meanwhile, the capacity of BMP to induce differentiation is constrained by LIF signaling. Thus, LIF and BMP signaling compensate each other and block the differentiation of mouse ES cells. (modified from [69])

3.2 BMP signaling

BMP signaling starts with phosphorylation of *SMAD1/5*. The phosphorylated *SMAD1/5* then combines with *SMAD4* and translocates into the nucleus to promote the transcription of inhibitor-of-differentiation (ID) genes (figure 1). In human ES cells, BMP signaling is inhibited in the undifferentiated state, while inducing trophoblast differentiation [30, 59, 85, 86]. The BMP antagonist, Noggin, as well as FGF2, play a repressive role on BMP signaling and sustain ES cell self-renewal in the absence of feeder cells or conditioned medium [81, 86] (figure 2). In mouse ES cells, combinatorial signaling of BMP and LIF is an important mechanism guiding pluripotency [17, 85, 87] (figure 1). The direction of mouse ES cell development seems dependent upon the levels of IDs and *MYC* in the nucleus. A constitutive expression of IDs can sustain prolonged self-renewal in the absence of BMP when cultured with a serum-free medium containing LIF [87]. In the absence of LIF, the inhibitory role of activated ID on differentiation is similar to the effect of BMP on mouse ES cells. Interestingly, the ID suppression along neural lineages compensates the *STAT3* inhibitory effect on differentiation along other lineages [87]. *STAT3* can also block BMP signaling. Apart from BMP, other pathways may induce the activation of ID and MYC expression, such as integrin pathways through fibronectin in serum, and the Notch pathway [5].

BMP signaling in combination with *extracellular receptor kinase (ERK)* and *P38 mitogen-activated protein kinase (MAPK)* also affect the fate of mouse ES cells (figures 1 and 2). By inhibiting both *ERK* and *P38 MAPK* signaling, BMP may be in favor of ES cell self-renewal [60]. The repression of *ERK* and *P38 MAPK* signaling mimics the effect of BMP in mouse ES cells. The *ERK* pathway can also be activated by LIF signaling, resulting in ES cell differentiation (figure 1). However, this process can be repressed by

BMP signaling, so preserving pluripotency (figure 1). The effect of BMP signaling on *ERK* and *P38 MAPK*, and whether this inhibition is related to ID activation, remains to be examined.

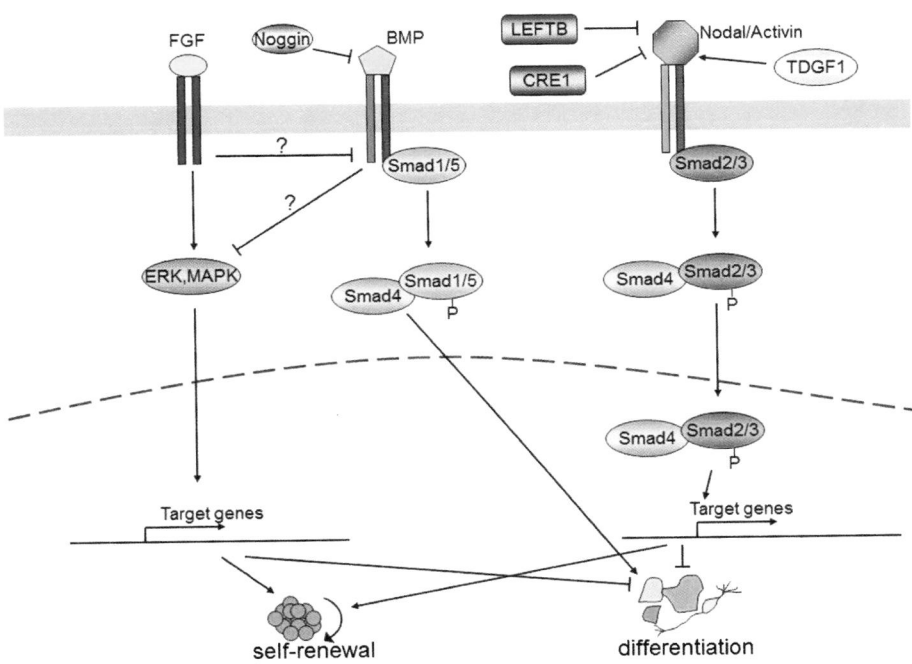

Figure 2. Collaboration between Activin/NODAL and FGF signaling maintains human ES cell pluripotency. Activin/NODAL signaling phosphorylate *SMAD2/3*, which associates with *SMAD4* and translocates into the nucleus. Target genes activated by *SMAD2/3* help maintain the pluripotent state of human ES cells but are not sufficient to completely prevent differentiation. Together, Activin/Nodal and FGF signaling can sustain prolonged proliferation of human ES cells in the absence of feeder-cells, conditioned medium, or serum replacer. BMP signaling promotes the differentiation of human ES cells and is repressed in undifferentiated human ES cells. The question marks denote possible interactions between FGF and BMP signaling pathways. (modified from [69])

3.3 Activin/NODAL and FGF signaling

Activin/NODAL and fibroblast growth factor (FGF) signaling are involved in maintenance of pluripotency in human ES cells. NODAL, like BMP, is also a member of the transforming growth factor-beta (TGFβ) superfamily. Activin/NODAL signaling is initiated by binding NODAL to the heterodimer of types I and II Activin receptors, which in turn phosphorylates *SMAD2/3* (figure 2) [63, 65]. The phosphorylated *SMAD2/3* forms a complex with *SMAD4*, and then the complex translocates into the nucleus to regulate the expression of target genes. Similarly, Activin, which is secreted from feeder cells used in ES cell culture, can also bind to the same Activin receptors and initiate the same *SMAD2/3* signaling. The co-factor *TDGF1* or antagonist *CRE1* modulates Activin/NODAL signaling through a feedback loop. The expression of *NODAL*, *LEFTA*, and *LEFTB* in undifferentiated human ES cells requires the activation of *SMAD2/3* [8].

The function of Activin/NODAL signaling in ES cell development is fulfilled through two alternative paths. In the first path, *SMAD2/3* is phosphorylated and localized in the nucleus to maintain human ES cells at an undifferentiated state. Once ES cells are differentiated, the phosphorylation and nuclear localization of *SMAD2/3* are greatly reduced [30]. In the second path, the Activin level is elevated to maintain pluripotency. It has been shown that Activin A maintains human ES cells in the undifferentiated state for a prolonged period of time in the absence of feeder layers, conditioned medium, or *STAT3* activation [7]. Conversely, the inhibition of Activin/NODAL signaling induces human ES cell differentiation [79]. However, neither NODAL nor Activin is enough to sustain a prolonged growth of human ES cell in a serum-free medium. FGF combined with either NODAL or

Activin can maintain pluripotency in the absence of feeder cells, conditioned medium, or serum replacer [79].

Besides interacting with BMP and LIF signaling, FGF signaling also targets *ERK* to prevent phosphorylated *SMAD* proteins from translocating into the nucleus [79]. However, FGF signaling inhibitors do not increase nuclear localization of phosphorylated *SMAD1/5* in human ES cells. It is reasoned that FGF signaling activates human ES cell self-renewal instead of blockage of BMP signaling. In mouse ES cells, however, FGF4 stimulation of *ERK* is an inductive stimulus that shifts ES cells toward lineage commitments [35].

3.4 WnT signaling

The canonical WnT signaling pathway functions within broad biological processes including cell differentiation, self-renewal, and proliferation. The binding of WnT proteins to the complex of Frizzled and LRP5/6 receptors at the cell surface triggers intracellular WnT/β-catenin signaling [3] (figure 3). This leads to a degradation of glycogen synthase kinase-3β (GSK3β) initiated by Dishevelled and FRAT. The deactivation of GSK3β causes an accumulation of β-catenin. The increased β-catenin in the cytoplasm then translocates into the nucleus and interacts with the complex consisting of TCF and lymphocyte enhancer factor (LEF), which are transcription factors enhancing the expression of WnT target genes, such as *MYC*, *TCF1*, and *Cyclin D1* [80].

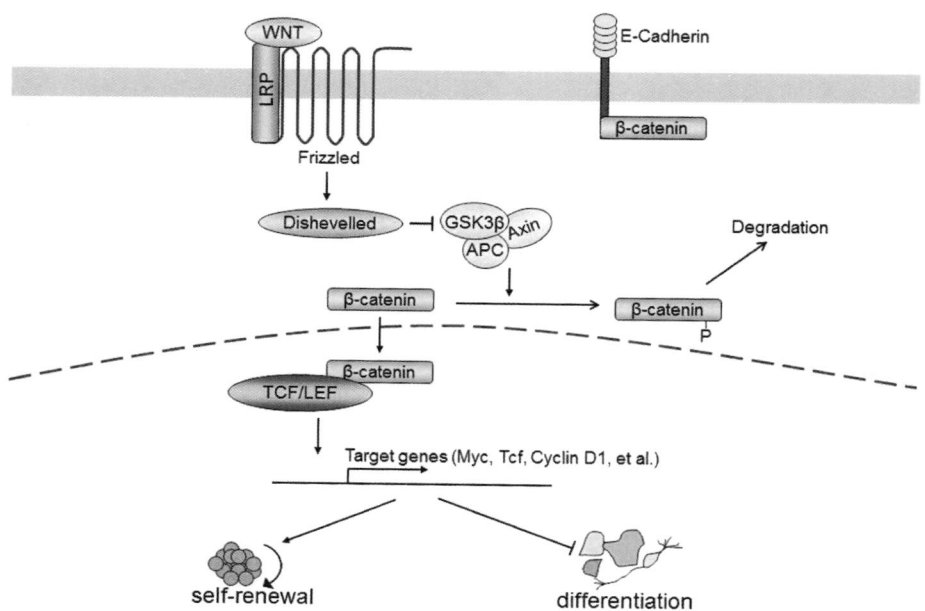

Figure 3. Canonical WnT signaling in maintaining pluripotency of human and mouse ES cells. Activation of Frizzled receptors by WnT ligands leads to the repression of GSK3β, thus preventing the degradation of β-catenin. The formation of the β-catenin/TCF complex activates transcription of target genes, which help to maintain the pluripotent state of ES cells. *E-Cadherin* can bind to β-catenin, which prevents its nuclear translocation and hence the β-catenin/TCF activated transcription. (modified from [69])

WnT signaling is important to the maintenance of pluripotency in both human and mouse ES cells [64]. In mouse ES cells, WnT3a represses GSK3β activity and increases *MYC* level, which favors self-renewal [64]. Upon removing Wn-T3a, ES cells undergo differentiation [14]. Applying GSK3β-specific inhibitors on human and mouse ES cells helps maintain the undifferentiated phenotypes and allows expression of pluripotent regulators *OCT4*, *REX1*, and *NANOG* [64]. After removing the inhibitors, ES cells undergo differentiation. WnT signaling also cooperates with LIF, extracellular matrix (ECM), or other pathways to maintain pluripotency [30].

MYC is a common marker linking both LIF and WnT signaling to maintain pluripotency, at least in mouse ES cells (Figures 1, 3). ECM-signaling stimulates the under-expression of *E-Cadherin* [53]. The inhibition of *E-Cadherin*, which binds to β-catenin and blocks its transportation to the nucleus, results in an accumulation of β-catenin and subsequent β-catenin/TCF mediated transcription, thus playing a similar function as triggered WnT signaling to maintain ES cell pluripotency [53].

4. Intrinsic Pluripotent TFs

TFs and related regulatory networks are the core facility to govern gene expression and mediate signaling pathways and cell processes in response to extracellular or developmental signals. Several TFs have been identified as master players that constitute the core regulatory architecture regulating ES cell pluripotency. The identification is particularly advanced by the chromatin immunoprecipitation (ChIP)-chip or ChIP-seq technologies [16, 33].

4.1 OCT4, SOX2, and NANOG

OCT4, *SOX2*, and *NANOG* are known as the core set of TFs maintaining ES cell pluripotency (see review articles of [11, 41, 49]). *OCT4* (or *POU5F1*) belongs to the POU TF family and is highly conserved on the genomic structure. Coordinating a variety of regulators in ES cells, *OCT4* is considered as the central pluripotent TF [11, 89]. A well-balanced level of *OCT4* expression is necessary to sustain ES cell pluripotency ([24, 28, 52]. An *OCT4*-deficient blastocyst fails to form ICM that generates pluripotent ES cells [50]. Knocking out *OCT4* in ES cells induces cell differentiation [24, 28].

SOX2 is a member of the *high mobility group (HMG-box)* family. Like OCT4, the deletion of SOX2 also results in the differentiation of ES cells, suggesting its importance for pluripotency. The determination of NANOG as an important pluripotent TF is based on its capacity to drive self-renewal of mouse ES cells in the absence of LIF and feeder layers [15, 47]. Similar to OCT4 and SOX2, NANOG is also evolutionarily conserved and highly expressed in both human and mouse ES cells [9, 82, 89]. Importantly, SOX2 and OCT4 form a heterodimer and further bind to NANOG. The three genes work collaboratively, forming the central regulatory circuitry in ES cells to determine the fate (figure 4) [12, 42, 81, 89]. The genes jointly activated by the three TFs are generally related to ES self-renewal, including TFs STAT3, ZIC3, and REST, as well as chromatin and histone modification factors SMARCAD1, MYST3, and SET [12, 42]. The genes repressed by the three TFs are responsible for ES cell differentiation, including HAND1, HOXB1, PAX6, and ISL1 [12]. On the other hand, OCT4, SOX2, and NANOG are able to stabilize the expression level of themselves through self-regulatory loops, as the SOX2-OCT4 heterodimer can bind to themselves and NANOG, while NANOG can bind to itself [12, 42, 89].

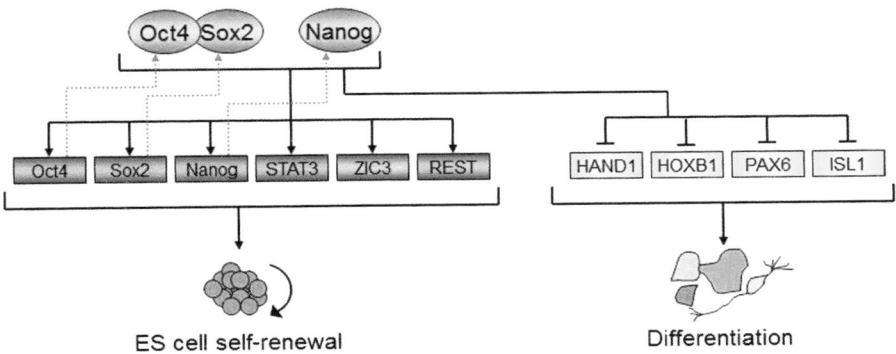

Figure 4. Oct4, Sox2, and Nanog constitute the core regulatory circuitry controlling ES cell pluripotency. Transcription factors are represented by ovals, and genes regulated by them are represented by rectangles. Selected genes activated (green) and repressed (yellow) by Oct4, Sox2, and Nanog are shown. Note that Oct4, Sox2, and Nanog regulate their own expression, forming auto-regulatory loops. (modified from [12]).

The OCT4/SOX2/NANOG network interacts with a number of signaling pathways in ES cells. The expression of FGF4 is regulated by OCT4 and SOX2 [79, 36, 89]. Two members of the NODAL pathway, TDGF1 and LEFTY2/EBAF, are also regulated by OCT4, SOX2, and NANOG [12, 89]. The overexpression of OCT4 and NANOG is measured in active LIF, BMP, Activin/NODAL, FGF, and WnT signaling pathways [30, 60, 64, 69, 84, 87]. The co-regulators or targets of OCT4/SOX2/NANOG, such as MYC, ESRRB, STAT3, TCF7, SALL4, FOXD3, FGF4, UTF1, and LRH-1 further interact with multiple pathways.

4.2 MYC and P53

MYC proteins, particularly c-MYC and n-MYC, are common effectors influencing critical signaling pathways in ES cells. They are nuclear proteins belonging to the family of helix-loop-helix/leucine zipper TFs. After heterodimer-ized with a small protein MAX, MYC activates or represses various target genes involved in cell proliferation, trans-formation, growth, differentiation, and apoptosis [2, 57]. In ES cells, MYC is linked functionally with LIF and WnT signaling and plays a role in maintaining pluripotency. Figure 5 shows the tight interactions among MYC, P53, and GSK3β. MYC expression can be regulated at two different levels in ES cells. First, the activation of STAT3 or β-catenin/TCF-LEF enhances the expression of MYC through both LIF and WnT pathways [14, 25]. Second, the LIF and WnT pathways inhibit GSK3β activity, thereby blocking T58 phosphorylation of MYC and increasing the half-life of MYC proteins. WnT signaling directly inhibits GSK3β activity, whereas LIF signaling represses its activ-ity by activating PI3K in ES cells [48, 54, 80]. PI3K, on the other hand, can be activated by insulin signaling or by ES-cell expressed Ras (Eras) [71], but inhibited by tumor suppressor phosphatase and tensin homologue deleted on chromosome 10 (PTEN) [72].

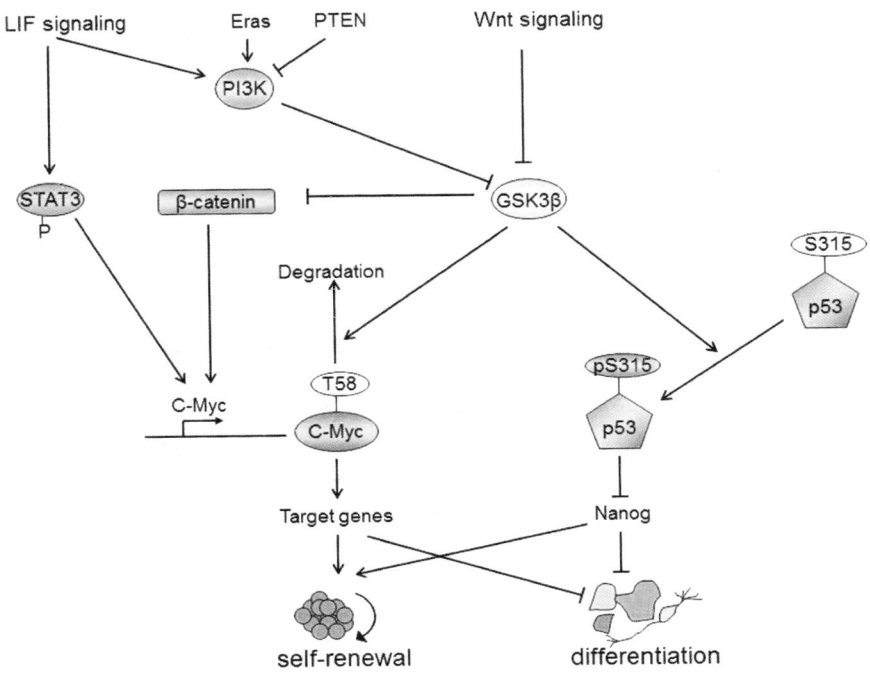

Figure 5. Interactions between GSK3β, MYC and P53 in ES cells. LIF and WnT signaling maintain an elevated level of MYC by activating MYC gene transcription and by inhibiting MYC protein degradation mediated by GSK3β. The elevated level of active MYC maintains ES cell pluripotency. In the absence of LIF and WnT, GSK3β is activated, and MYC is phosphorylated at T58, inducing its degradation. Active GSK3β also activates P53 by phosphorylation at S315, leading to the suppression of Nanog expression and promoting ES cell differentiation. (modified from [69])

P53 is a top tumor suppressor and induces cell cycle arrest and apoptosis in response to DNA damage [6]. In mouse ES cells, P53-induced cell cycle arrest is restored after ES cells undergo differentiation [1]. Its activation by the phosphorylation at Ser 315 inhibits the expression of Nanog and stimulates ES cell differentiation [40]. Ser 315 of P53 is also a substrate of GSK3β [61, 136]. The suppressive role of P53 on mouse ES cells may connect to other pathways like WnT, LIF, and PI3K through GSK3β (figure 5). Multiple pathways work together through the dynamic change of GSK3β activity which regulates *MYC* and *P53*.

5. Epigenetic Regulation

Beside transcriptional regulation, epigenetic regulation also plays a significant role in ES cell self-renewal and differentiation. Several epigenetic modifications, including histone modifications, DNA methylation, and non-coding RNA-mediated regulation, have been particularly investigated for their roles in regulating ES cell pluripotency.

5.1 Histone modification

Chromatin is a complex consisting of DNA, histones, and other proteins that form chromosomes. In pluripotent ES cells, chromatin structure is dynamic and loosely bound and remains in an active state, open for transcriptional regulation. During ES cell differentiation, chromatin structure becomes stable, condensed, and at an inactive state. Hence, chromatin is a center for epigenetic regulation. Histone modification such as methylation, acetylation, or phosphorylation is able to alter chromatin structure so as to specify the lineage and developmental stage of ES cells through epigenetic regulation on gene expression. It has been shown that the promoters of pluripotent genes are concurrently marked by activated methylations or 3 lysine 4 trimethylation (H3K4me3) of histones, whereas genes associated with differentiation are enriched with repressive ones (H3K-27me3) [4]. This bivalent pattern of histone modifications helps to specify ES cell identity. H3K4me3 is associated with cell proliferation and likely related to ES cell self-renewal

[90]. The promoters of *OCT4*, *SOX2*, and *NANOG* are highly occupied by H3K4me3 [55, 90]. H3K27me3, on the other hand, is associated with gene silencing [55] and processed by *polycomb repressor complex 2 (PRC2)*. *PRC2* components are critical for early embryo development [13], and their target genes are greatly enriched by TFs crucial to ES cell development and over-expressed upon induction of ES cell differentiation [13, 38].

Using high-throughput quantitative chromatin profiling, *NANOG* and several other pluripotency regulators (*GDF3*, *DPPA3*) are identified with DNaseI hypersensitive sites (HS) along a 160 kb region of chromosome 6 where they reside [39]. These pluripotency factors plus *OCT4* form a chromosomal bridge within the HS sites by altering the higher order chromatin structure in pluripotent ES cells. *OCT4* expression is required to stabilize this structure, as the depletion of *OCT4* destroys such chromatin organization and simultaneously loses ES cell pluripotency [39].

5.2 DNA methylation

DNA methylation is a general indicator of silenced genome regions, essential for the establishment of chromatin structure during development [37, 43]. ES cell-specific genes have been examined for the epigenetic status, and correlations between ES cell development and DNA methylation have been established [10, 23, 32, 44]. Most of methylated genes are associated with differentiation and signal transduction in mouse ES cells [20]. Unmethylated genes are largely enriched with transcriptional activities and cell survival. They are repressed by polycomb complex proteins or targeted by *OCT4* and *NANOG*, possibly involved in ES cell self-renewal [20].

5.3 MicroRNAs

MicroRNAs represent 20–25 nucleotide non-coding RNAs that bind to target mRNAs influencing their translation and stabilization. MicroRNAs regulate ES cell self-renewal and differentiation through post-transcriptional repression of the translation of target mRNA in ES cells [21]. It has been shown that Dicer-deficient mouse ES cells, which cannot generate microRNAs, are defective in differentiation into three germ layer tissues following LIF withdrawal [31]. The observation indicates that microRNAs play a critical role in controlling ES cell differentiation. ES cell–specific microRNAs have been identified in both human and mouse cells and shown to play a specific role for ES cell self-renewal and early differentiation [27,67]. These microRNAs are expressed only in undifferentiated ES cells, but their expression is significantly reduced once ES cells differentiate into embryoid bodies, and becomes undetectable in adult organs [27, 67]. MicroRNAs preferably expressed in other tissues or organs are poorly or not expressed in ES cells [27, 67]. ES cell–specific microRNAs may interact with core pluripotent transcriptional regulators, and they act in a complementary role to these regulators for demarcating ES cell potential during development. It has been shown that miR-134, miR-296, and miR-470 co-target coding regions of *Oct4*, *Sox2*, and *Nanog* in a variety of combinations, resulting in differentiation of mouse ES cells as they are over-expressed upon differentiation [75]. miR-21 contains potential binding sites that target *Oct4*, *Sox2*, and *Nanog*. An elevated level of miR-21 can specifically suppress the self-renewal of mouse ES cells, corresponding to the under-expression of *Oct4*, *Sox2*, *Nanog*, and c-MYC. A study based on genome-wide ChIP-seq technology indicates that the promoters of ~20% ES cell–specific microRNA genes co-localize with binding sites of *Oct4*, *Sox2*, *Nanog*, and *TCF3* [45]. A pluripotency control by concurrent regulation of cell cycle under collaboration between transcriptional networks and microRNAs is also demonstrated [11]. MicroRNAs that are activated in ES cells by *Oct4/Sox2/Nanog* and other factors may help modulate the direct effects of these TFs by acting on their common target genes, and determine ES cells for efficient differentiation.

6. Induced Pluripotent Stem Cell

One of the remarkable scientific achievements in recent stem cell research is the generation of induced pluripotent (iPS) cells from somatic cells. iPS cells are believed to be identical to ES cells in many respects, such as the expression of signature genes and proteins, the chromatin methylation pattern, and the formation of EBs, teratoma, and viable chimera, in addition to the potency and differentiation ability [73, 88]. iPS cells are firstly generated from adult murine fibroblasts by forced expression of the TFs *Oct4*, *Sox2*, *Klf4*, and *c-MYC* [74]. Later, iPS cells are generated from human adult cells using the same four TFs or by replacing *Klf4* and *c-MYC* with *Nanog* and *Lin28* [73, 88]. Recent results demonstrate that *Oct4* is an indispensable reprogramming factor, and only two factors, *Oct4* and *Klf4*, are requisite to reprogram adult neural stem cells for iPS cells; while *MYC* may not be necessary for the recapture of pluripotency [34, 88]. In addition to TFs, epigenetic factors are additional candidates for functional reprogramming to induce pluripotency. A recent review article summarizes a list of potential molecules which can be utilized to produce iPS cells by reprogramming somatic cells [19]. These molecules include genetic factors, chemical inhibitors, and signaling molecules that can substitute for known iPS inducing TFs. Although various iPS inducing factors have been identified, it remains

to be determined how they work coordinately to reprogram cells.

7. Evolutionary Conservation of Mechanisms Controlling Pluripotency

Despite the common features, significant differences exist among ES cells isolated from different species [22, 26, 66, 77]. Cross-species comparative analysis of ES cells allows identifying fundamental and species-specific mechanisms regulating ES cell pluripotency. A human-mouse compara-tive genomics study demonstrates that the pathways di-rected by *Oct4*, *Sox2*, *Nanog*, and *Nodal* are evolutionarily conserved as the component genes are conserved on both the gene and promoter structure [89]. The LIF pathway is, on the other hand, evolutionarily divergent from the genomic perspective. A human-mouse comparative transcriptomics study examined a large set of biological pathways and pro-cesses, as well as transcription factors and growth factors expressed in ES cells [70]. Some of the biological processes or pathways are found to be conserved as the genes show a significant cross-species correlation on the transcriptional response to ES cell differentiation. Many of the conserved biological processes are involved in development, includ-ing cellular morphogenesis, embryonic development, pat-tern specification, and determination of left-right symmetry (figure 6).

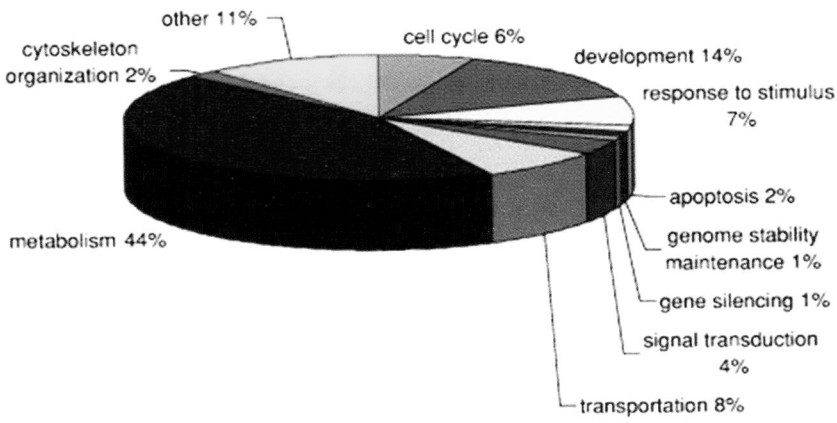

Figure 6. Major categories of biological processes with conserved expression patterns between human and mouse ES cells. (modified from [70])

The conserved biological processes also include establishment and/or maintenance of chromatin architecture, responses to extrinsic stimuli, cell proliferation and cell cycle, signal transduction, apoptosis, and TGF-β, WnT, Nodal, and ingerin-mediated signaling pathways. Transcription or growth factors such as *GDF3*, *LEFTB*, *MYB*, *MYCN*, *NFYB*, *POLR3K*, *POU2F1*, *TDFG1*, and *UTF1* are also conserved in the transcriptional response to ES cell differentiation. These conserved pathways and factors represent fundamental molecular mechanisms regulating ES cell pluripotency. On the other hand, some biological processes or pathways are not conserved as showing negative or no correlation in the expression patterns across species [70]. For example, members of the LIF pathway show divergent expression profiles across species, consistent with the observation that LIF signaling is critical for mouse ES cells but not functional in human ES cells [17, 29]. Genes involved in cytokine-cytokine receptor interactions also show divergent expression profiles, suggestive of distinct responses of human and mouse

ES cells to those growth factors. Moreover, members of the FGF pathway show divergent transcriptional patterns, consistent with the observation that FGF4 shows a sequence divergence on a *Sox2/Oct4* co-binding site that is critical for ES cell–specific expression [36, 89]. The evolutionarily divergent pathways are indicative of species-specific mechanisms controlling ES cell self-renewal and differentiation. A human-mouse comparative study also examined gene co-expression in each ES cell–critical pathway [68]. The study identified complexes of co-expressed genes that are conserved across species or unique to a single species and thus indicative of fundamental or species-specific modulation of gene expression controlling ES cell pluripotency. The results suggest an essential role of *JAK*-mediated signaling through activating *STAT2* and *PI3K* in ES cells, while reaffirming different requirements of STAT3–mediated LIF signaling in mouse and human ES cells. The mechanisms of WnT signaling seem to be different in human and mouse ES cells as the pathway shows differential co-expression among the key component genes across species. The AKT/PTEN pathway shows a high co-expression among the key members in both species, suggesting its fundamental role in ES cell development. A set of transcription factors, including FOX, GATA, MYB, NANOG, OCT, PAX, SOX, and STAT, as well the FGF response element, may underlie the conserved co-expression in the ES cell–critical pathways as their binding sites are significantly enriched in the promoter sequences of the pathway genes. A global co-expression network conserved between human and mouse ES cells offers further insights into transcriptional regulation in ES cells [68]. The network is dominated by a few highly-connected genes (hub genes) that link the less-connected genes to the system. The hub genes, including *IGF2, JARID2, LCK, MYCN, NASP, OCT4, ORC1L, PHC1* and *RUVBL1*, are possibly critical in determining the fate of ES cells.

8. Summary

ES cells either maintain the pluripotent state by self-renewal or undergo differentiation. By integrating data from different sources, a global picture is emerging as to how ES cells determine their fate. Under conditions favoring ES cell self-renewal, the combinatorial effects of key signaling pathways, such as LIF, BMP, WnT, and others, allow the up-regulation of *Oct4*, *Sox2*, and *Nanog*, which in turn activate the expression of genes critical for long-term self-renewal and repress the expression of genes initiating differentiation. Feedback regulation of this regulatory core helps maintain the stability of the network. GSK3β is a common target of LIF, WnT, IGF, and ERK pathways and facilitates the interaction between extracellular signals and intracellular pathways in ES cells. The activity of GSK3β can be repressed by combinatorial effects of different signaling pathways, resulting in an elevated level of active *MYC* and the repression of *P53*. Following the withdrawal of critical growth factors, the suppression of GSK3β is lifted, resulting in a rapid reduction of active *MYC* and the activation of *P53*, which represses the expression of *Nanog*, *Oct4*, and Sox2 and commits the ES cells into differentiation. Additional regulatory mechanisms related to microRNAs, epigenetic regulation, and genome domain organization may also play important roles in maintaining ES cell pluripotency. Despite the fact that ES cells from different organisms respond differently to different growth factors, the *Oct4*/*Sox2*/*Nanog* core regulatory network is evolutionarily conserved. Fundamental or species-specific mechanisms controlling ES cell pluripotency can be elucidated from evolutionarily conserved or divergent patterns at genomic, transcriptomic, or network levels by cross-species comparative analysis of ES cells.

References

1. Aladjem, M. I., Spike, B. T., Rodewald, L. W., Hope, T. J., Klemm, M., Jaenisch, R., and Wahl, G. M. 1998. ES cells do not activate p53-dependent stress responses and undergo p53-independent apoptosis in response to DNA damage. *Curr Biol* 8:145–155.

2. Amati, B., Frank, S. R., Donjerkovic, D., and Taubert, S. 2001. Function of the c-MYC oncoprotein in chromatin remodeling and transcription. *Biochim Biophys Acta* 1471: M135–145.

3. Angers, S. and Moon, R. T. 2009. Proximal events in WnT signal transduction. *Nat Rev Mol Cell Biol* 10:468–477.

4. Azuara, V., Perry, P., Sauer, S., Spivakov, M., Jorgensen, H. F., John, R. M., Gouti, M., Casanova, M., Warnes, G., Merkenschlager, M., and Fisher, A.G. 2006. Chromatin signatures of pluripotent cell lines. *Nat Cell Biol* 8:532–538.

5. Baonza, A., de Celis, J. F., and Garcia-Bellido, A. 2000. Relationships between extramacrochaetae and Notch signalling in Drosophila wing development. *Development* 127:2383–2393.

6. Bargonetti, J. and Manfredi, J. J. 2002. Multiple roles of the tumor suppressor p53. *Curr Opin Oncol* 14:86–91.

7. Beattie, G. M., Lopez, A. D., Bucay, N., Hinton, A., Firpo, M. T., King, C. C., and Hayek, A. 2005. Activin A maintains pluripotency of human embryonic stem cells in the absence of feeder layers. *Stem Cells* 23:489–495.

8. Besser, D. 2004. Expression of nodal, lefty-a, and lefty-B in undifferentiated human embryonic stem cells requires activation of SMAD2/3. *J Biol Chem* 279:45076–45084.

9. Bhattacharya, B., Miura, T., Brandenberger, R., Mejido, J., Luo, Y., Yang, A. X., Joshi, B. H., Ginis, I., Thies, R. S., Amit, M., Lyons, I., Condie, B. G., Itskovitz-Eldor, J., Rao, M. S., and Puri, R.K. 2004. Gene expression in human embryonic stem cell lines: unique molecular signature. *Blood* 103:2956–2964.

10. Bibikova, M., Chudin, E., Wu, B., Zhou, L., Garcia, E. W., Liu, Y., Shin, S., Plaia, T. W., Auerbach, J. M., Arking, D. E., Gonzalez, R., Crook, J., Davidson, B., Schulz, T. C., Robins, A., Khanna, A., Sartipy, P., Hyllner, J., Vanguri, P., Savant-Bhonsale, S., Smith, A. K., Chakravarti, A., Maitra, A., Rao, M., Barker, D. L., Loring, J. F., and Fan, J. B. 2006. Human embryonic stem cells have a unique epigenetic signature. *Genome Res* 16:1075–1083.

11. Bosnali, M., Munst, B., Thier, M., and Edenhofer, F. 2009. Deciphering the stem cell machinery as a basis for understanding the molecular mechanism underlying reprogramming. *Cell Mol LIFe Sci.*

12. Boyer, L. A., Lee, T. I., Cole, M. F., Johnstone, S. E., Levine, S. S., Zucker, J. P., Guenther, M. G., Kumar, R. M., Murray, H. L., Jenner, R. G., Gifford, D. K., Melton, D. A., Jaenisch, R., and Young, R.A. 2005. Core transcriptional regulatory circuitry in human embryonic stem cells. *Cell* 122:947–956.

13. Boyer, L. A., Plath, K., Zeitlinger, J., Brambrink, T., Medeiros, L. A., Lee, T. I., Levine, S. S., Wernig, M., Tajonar, A., Ray, M. K., Bell, G. W., Otte, A. P., Vidal, M., Gifford, D. K., Young, R. A., and Jaenisch, R. 2006. Polycomb complexes repress developmental regulators in murine embryonic stem cells. *Nature* 441:349–353.

14. Cartwright, P., McLean, C., Sheppard, A., Rivett, D., Jones, K., and Dalton, S. 2005. LIF/STAT3 controls ES cell self-renew-

al and pluripotency by a MYC-dependent mechanism. *Development* 132:885–896.

15. Chambers, I., Colby, D., Robertson, M., Nichols, J., Lee, S., Tweedie, S., and Smith, A. 2003. Functional expression cloning of Nanog, a pluripotency sustaining factor in embryonic stem cells. *Cell* 113:643–655.

16. Chen, X., Xu, H., Yuan, P., Fang, F., Huss, M., Vega, V.B., Wong, E., Orlov, Y. L., Zhang, W., Jiang, J., Loh, Y. H., Yeo, H. C., Yeo, Z. X., Narang, V., Govindarajan, K. R., Leong, B., Shahab, A., Ruan, Y., Bourque, G., Sung, W. K., Clarke, N. D., Wei, C. L., and Ng, H. H. 2008. Integration of external signaling pathways with the core transcriptional network in embryonic stem cells. *Cell* 133:1106–1117.

17. Daheron, L., Opitz, S. L., Zaehres, H., Lensch, W. M., Andrews, P. W., Itskovitz-Eldor, J., and Daley, G.Q. 2004. LIF/*STAT3* signaling fails to maintain self-renewal of human embryonic stem cells. *Stem Cells* 22:770–778.

18. Evans, M. J. and Kaufman, M. H. 1981. Establishment in culture of pluripotential cells from mouse embryos. *Nature* 292:154–156.

19. Feng, B., Ng, J. H., Heng, J. C., and Ng, H. H. 2009. Molecules that promote or enhance reprogramming of somatic cells to induced pluripotent stem cells. *Cell Stem Cell* 4:301–312.

20. Fouse, S. D., Shen, Y., Pellegrini, M., Cole, S., Meissner, A., Van Neste, L., Jaenisch, R., and Fan, G. 2008. Promoter CpG methylation contributes to ES cell gene regulation in parallel with *Oct4*/Nanog, PcG complex, and histone H3 K4/K27 trimethylation. *Cell Stem Cell* 2:160–169.

21. Gangaraju, V. K. and Lin, H. 2009. MicroRNAs: key regulators of stem cells. *Nat Rev Mol Cell Biol* 10:116–125.

22. Ginis, I., Luo, Y., Miura, T., Thies, S., Brandenberger, R., Gerecht-Nir, S., Amit, M., Hoke, A., Carpenter, M. K., Itskovitz-Eldor, J., and Rao, M. S. 2004. Differences between human and mouse embryonic stem cells. *Dev Biol* 269:360–380.

23. Hattori, N., Nishino, K., Ko, Y. G., Hattori, N., Ohgane, J., Tanaka, S., and Shiota, K. 2004. Epigenetic control of mouse *Oct4* gene expression in embryonic stem cells and trophoblast stem cells. *J Biol Chem* 279:17063–17069.

24. Hay, D. C., Sutherland, L., Clark, J., and Burdon, T. 2004. *Oct4* knockdown induces similar patterns of endoderm and trophoblast differentiation markers in human and mouse embryonic stem cells. *Stem Cells* 22:25–235.

25. He, T. C., Sparks, A. B., Rago, C., Hermeking, H., Zawel, L., da Costa, L. T., Morin, P. J., Vogelstein, B., and Kinzler, K.W. 1998. Identification of c-MYC as a target of the APC pathway, *Science*. 281:1509–1512.

26. Hong, Y., Winkler, C., and Schartl, M. 1996. Pluripotency and differentiation of embryonic stem cell lines from the medakafish Oryzias latipes., *Mech Dev* 60:33–44.

27. Houbaviy, H. B., Murray, M. F., and Sharp, P. A. 2003. Embryonic stem cell-specific MicroRNAs. *Dev Cell* 5:351–358.

28. Hough, S. R., Clements, I., Welch, P. J., and Wiederholt, K. A. 2006. Differentiation of mouse embryonic stem cells after RNA interference-mediated silencing of *OCT4* and Nanog. *Stem Cells* 24:1467–1475.

29. Humphrey, R. K., Beattie, G. M., Lopez, A. D., Bucay, N., King, C. C., Firpo, M. T., Rose-John, S., and Hayek, A. 2004. Maintenance of pluripotency in human embryonic stem cells is STAT3 independent. *Stem Cells* 22:522–530.

30. James, D., Levine, A. J., Besser, D., and Hemmati-Brivanlou, A. 2005. TGFbeta/activin/nodal signaling is necessary for the maintenance of pluripotency in human embryonic stem cells. *Development* 132:1273–1282.

31. Kanellopoulou, C., Muljo, S. A., Kung, A. L., Ganesan, S., Drapkin, R., Jenuwein, T., Livingston, D. M., and Rajewsky, K. 2005. Dicer-deficient mouse embryonic stem cells are defective in differentiation and centromeric silencing, *Genes Dev* 19:489–501.

32. Keller, G. 2005. Embryonic stem cell differentiation: emergence of a new era in biology and medicine. *Genes Dev* 19:1129–1155.

33. Kim, J., Chu, J., Shen, X., Wang, J., and Orkin, S. H. 2008. An extended transcriptional network for pluripotency of embryonic stem cells, *Cell* 132:1049–1061.

34. Kim, J. B., Zaehres, H., Wu, G., Gentile, L., Ko, K., Sebastiano, V., Arauzo-Bravo, M. J., Ruau, D., Han, D. W., Zenke, M., and Scholer, H. R. 2008. Pluripotent stem cells induced from adult neural stem cells by reprogramming with two factors. *Nature* 454:646–650.

35. Kunath, T., Saba-El-Leil, M. K., Almousailleakh, M., Wray, J., Meloche, S., and Smith, A. 2007. FGF stimulation of the Erk1/2 signalling cascade triggers transition of pluripotent embryonic stem cells from self-renewal to lineage commitment. *Development* 134:2895–2902.

36. Lamb, K. A. and Rizzino, A. 1998. Effects of differentiation on the transcriptional regulation of the FGF-4 gene: critical roles played by a distal enhancer. *Mol Reprod Dev* 51:218–224.

37. Lande-Diner, L. and Cedar, H. 2005. Silence of the genes-- mechanisms of long-term repression. *Nat Rev Genet* 6:648–654.

38. Lee, T. I., Jenner, R. G., Boyer, L. A., Guenther, M. G., Levine, S. S., Kumar, R. M., Chevalier, B., Johnstone, S. E., Cole, M. F., Isono, K., Koseki, H., Fuchikami, T., Abe, K., Murray, H. L., Zucker, J. P., Yuan, B., Bell, G. W., Herbolsheimer, E., Hannett, N. M., Sun, K., Odom, D. T., Otte, A. P., Volkert, T. L., Bartel, D. P., Melton, D. A., Gifford, D. K., Jaenisch, R., and Young, R.A. 2006. Control of developmental regulators by Polycomb in human embryonic stem cells, *Cell* 125:301–313.

39. Levasseur, D. N., Wang, J., Dorschner, M. O., Stamatoyan-nopoulos, J. A. and Orkin, S. H. 2008. *Oct4* dependence of chromatin structure within the extended Nanog locus in ES cells. *Genes Dev* 22:575–580.

40. Lin, T., Chao, C., Saito, S., Mazur, S. J., Murphy, M. E., Appel-la, E., and Xu, Y. 2005. p53 induces differentiation of mouse embryonic stem cells by suppressing Nanog expression. *Nat Cell Biol* 7:165–171.

41. Loh, Y. H., Ng, J. H., and Ng, H. H. 2008. Molecular framework underlying pluripotency. *Cell Cycle* 7:885–891.

42. Loh, Y. H., Wu, Q., Chew, J. L., Vega, V. B., Zhang, W., Chen, X., Bourque, G., George, J., Leong, B., Liu, J., Wong, K. Y., Sung, K. W., Lee, C. W., Zhao, X. D., Chiu, K. P., Lipovich, L., Kuznetsov, V. A., Robson, P., Stanton, L. W., Wei, C. L., Ruan, Y., Lim, B., and Ng, H. H. 2006. The *Oct4* and Nanog transcrip-

tion network regulates pluripotency in mouse embryonic stem cells. *Nat Genet* 38:431–440.

43. Lorincz, M. C., Dickerson, D. R., Schmitt, M., and Groudine, M. 2004. Intragenic DNA methylation alters chromatin structure and elongation efficiency in mammalian cells. *Nat Struct Mol Biol* 11:1068–1075.

44. Maitra, A., Arking, D. E., Shivapurkar, N., Ikeda, M., Stastny, V., Kassauei, K., Sui, G., Cutler, D. J., Liu, Y., Brimble, S. N., Noaksson, K., Hyllner, J., Schulz, T.C., Zeng, X., Freed, W. J., Crook, J., Abraham, S., Colman, A., Sartipy, P., Matsui, S., Carpenter, M., Gazdar, A. F., Rao, M., and Chakravarti, A. 2005. Genomic alterations in cultured human embryonic stem cells. *Nat Genet* 37:1099–1103.

45. Marson, A., Levine, S. S., Cole, M. F., Frampton, G. M., Brambrink, T., Johnstone, S., Guenther, M. G., Johnston, W. K., Wernig, M., Newman, J., Calabrese, J. M., Dennis, L. M., Volkert, T. L., Gupta, S., Love, J., Hannett, N., Sharp, P. A., Bartel, D. P., Jaenisch, R., and Young, R. A. 2008. Connecting microRNA genes to the core transcriptional regulatory circuitry of embryonic stem cells, *Cell* 134:521–533.

46. Martin, G.R. 1981. Isolation of a pluripotent cell line from early mouse embryos cultured in medium conditioned by teratocarcinoma stem cells. *Proc Natl Acad Sci USA* 78:7634–7638.

47. Mitsui, K., Tokuzawa, Y., Itoh, H., Segawa, K., Murakami, M., Takahashi, K., Maruyama, M., Maeda, M., and Yamanaka, S. 2003. The homeoprotein Nanog is required for maintenance of pluripotency in mouse epiblast and ES cells. *Cell* 113:631–642.

48. Moon, R. T., Bowerman, B., Boutros, M., and Perrimon, N. 2002. The promise and perils of WnT signaling through beta-catenin. *Science* 296:1644–1646.

49. Ng, J. H., Heng, J. C., Loh, Y. H., and Ng, H. H. 2008. Transcriptional and epigenetic regulations of embryonic stem cells. *Mutat Res* 647:52–58.

50. Nichols, J., Zevnik, B., Anastassiadis, K., Niwa, H., Klewe-Nebenius, D., Chambers, I., Scholer, H., and Smith, A. 1998. Formation of pluripotent stem cells in the mammalian embryo depends on the POU transcription factor *Oct4*, *Cell* 95:379–391.

51. Niwa, H. 2001. Molecular mechanism to maintain stem cell renewal of ES cells. *Cell Struct Funct* 26:137–148.

52. Niwa, H., Miyazaki, J., and Smith, A. G. 2000. Quantitative expression of Oct-3/4 defines differentiation, dedifferentiation or self-renewal of ES cells. *Nat Genet* 24:372–376.

53. Oloumi, A., McPhee, T., and Dedhar, S. 2004. Regulation of E-cadherin expression and beta-catenin/Tcf transcriptional activity by the integrin-linked kinase. *Biochim Biophys Acta* 1691:1–15.

54. Paling, N. R., Wheadon, H., Bone, H. K., and Welham, M.J. 2004. Regulation of embryonic stem cell self-renewal by phosphoinositide 3-kinase-dependent signaling. *J Biol Chem* 279:48063–48070.

55. Pan, G., Tian, S., Nie, J., Yang, C., Ruotti, V., Wei, H., Jonsdottir, G. A., Stewart, R., and Thomson, J. A. 2007. Whole-genome analysis of histone H3 lysine 4 and lysine 27 methylation in human embryonic stem cells. *Cell Stem Cell* 1:299–312.

56. Park, I. H., Zhao, R., West, J. A., Yabuuchi, A., Huo, H., Ince, T. A., Lerou, P. H., Lensch, M. W., and Daley, G. Q. 2008. Reprogramming of human somatic cells to pluripotency with defined factors. *Nature* 451:141–146.

57. Patel, J. H., Loboda, A. P., Showe, M. K., Showe, L. C., and McMahon, S. B. 2004. Analysis of genomic targets reveals complex functions of MYC. *Nat Rev Cancer* 4:562–568.

58. Pei, D. 2009. Regulation of pluripotency and reprogramming by transcription factors. *J Biol Chem* 284:3365–3369.

59. Pera, M. F., Andrade, J., Houssami, S., Reubinoff, B., Trounson, A., Stanley, E. G., Ward-van Oostwaard, D., and Mummery, C. 2004. Regulation of human embryonic stem cell differentiation by BMP-2 and its antagonist noggin. *J Cell Sci* 117:1269–1280.

60. Qi, X., Li, T. G., Hao, J., Hu, J., Wang, J., Simmons, H., Miura, S., Mishina, Y., and Zhao, G. Q. 2004. BMP4 supports self-renewal of embryonic stem cells by inhibiting mitogen-activated protein kinase pathways. *Proc Natl Acad Sci USA* 101:6027–6032.

61. Qu, L., Huang, S., Baltzis, D., Rivas-Estilla, A. M., Pluquet, O., Hatzoglou, M., Koumenis, C., Taya, Y., Yoshimura, A., and Koromilas, A. E. 2004. Endoplasmic reticulum stress induces p53 cytoplasmic localization and prevents p53-dependent apoptosis by a pathway involving glycogen synthase kinase-3beta. *Genes Dev* 18:261–277.

62. Rao, M. 2004. Conserved and divergent paths that regulate self-renewal in mouse and human embryonic stem cells. *Dev Biol* 275:269–286.

63. Reissmann, E., Jornvall, H., Blokzijl, A., Andersson, O., Chang, C., Minchiotti, G., Persico, M. G., Ibanez, C. F., and Brivan-lou, A.H. 2001. The orphan receptor ALK7 and the Activin receptor ALK4 mediate signaling by Nodal proteins during vertebrate development. *Genes Dev* 15:2010–2022.

64. Sato, N., Meijer, L., Skaltsounis, L., Greengard, P., and Brivan-lou, A.H. 2004. Maintenance of pluripotency in human and mouse embryonic stem cells through activation of WnT sig-naling by a pharmacological GSK-3-specific inhibitor. *Nat Med* 10:55–63.

65. Schier, A. F. 2003. Nodal signaling in vertebrate develop-ment. *Annu Rev Cell Dev Biol* 19:589–621.

66. Soodeen-Karamath, S. and Gibbins, A. M. 2001. Apparent absence of Oct 3/4 from the chicken genome. *Mol Reprod Dev* 58:137–148.

67. Suh, M. R., Lee, Y., Kim, J. Y., Kim, S. K., Moon, S. H., Lee, J. Y., Cha, K. Y., Chung, H. M., Yoon, H. S., Moon, S. Y., Kim, V. N., and Kim, K. S. 2004. Human embryonic stem cells express a unique set of microRNAs. *Dev Biol* 270:488–498.

68. Sun, Y., Li, H., Liu, Y., Mattson, M. P., Rao, M. S., and Zhan, M. 2008. Evolutionarily conserved transcriptional co-expres-sion guiding embryonic stem cell differentiation. *PLoS ONE* 3:e3406.

69. Sun, Y., Li, H., Liu, Y., Shin, S., Mattson, M. P., Rao, M., and Zhan, M. 2006. Mechanisms controlling embryonic stem cell self-renewal and differentiation. *Critical Rev Mammal Gene Expr* 16:198.

70. Sun, Y., Li, H., Liu, Y., Shin, S., Mattson, M. P., Rao, M. S., and Zhan, M. 2007. Cross-species transcriptional profiles establish

a functional portrait of embryonic stem cells. *Genomics* 89:22–35.

71. Takahashi, K., Mitsui, K., and Yamanaka, S. 2003. Role of ERas in promoting tumour-like properties in mouse embryonic stem cells. *Nature* 423:541–545.

72. Takahashi, K., Murakami, M., and Yamanaka, S. 2005. Role of the phosphoinositide 3-kinase pathway in mouse embryonic stem ES. cells. *Biochem Soc Trans* 33:1522–1525.

73. Takahashi, K., Tanabe, K., Ohnuki, M., Narita, M., Ichisaka, T., Tomoda, K., and Yamanaka, S. 2007. Induction of pluripotent stem cells from adult human fibroblasts by defined factors. *Cell* 131:861–872.

74. Takahashi, K. and Yamanaka, S. 2006. Induction of pluripotent stem cells from mouse embryonic and adult fibroblast cultures by defined factors. *Cell* 126:663–676.

75. Tay, Y., Zhang, J., Thomson, A. M., Lim, B., and Rigoutsos, I. 2008. MicroRNAs to Nanog, *Oct4* and Sox2 coding regions modulate embryonic stem cell differentiation. *Nature* 455:1124–1128.

76. Thomson, J. A., Itskovitz-Eldor, J., Shapiro, S. S., Waknitz, M. A., Swiergiel, J. J., Marshall, V. S., and Jones, J. M. 1998. Embryonic Stem Cell Lines Derived from Human Blastocysts. *Science* 282:1145–1147.

77. Thomson, J. A., Kalishman, J., Golos, T. G., Durning, M., Harris, C. P., Becker, R. A., and Hearn, J. P. 1995. Isolation of a primate embryonic stem cell line. *Proc Natl Acad Sci USA* 92:7844–7848.

78. Thomson, J. A., Kalishman, J., Golos, T. G., Durning, M., Harris, C. P., and Hearn, J. P. 1996. Pluripotent cell lines derived from common marmoset Callithrix jacchus. blastocysts. *Biol Reprod* 55:254–259.

79. Vallier, L., Alexander, M., and Pedersen, R.A. 2005. Activin/ Nodal and FGF pathways cooperate to maintain pluripotency of human embryonic stem cells. *J Cell Sci* 118:4495–4509.

80. Van Es, J. H., Barker, N., and Clevers, H. 2003. You WnT some, you lose some: oncogenes in the WnT signaling pathway. *Curr Opin Genet Dev* 13:28–33.

81. Wang, G., Zhang, H., Zhao, Y., Li, J., Cai, J., Wang, P., Meng, S., Feng, J., Miao, C., Ding, M., Li, D., and Deng, H. 2005. Noggin and bFGF cooperate to maintain the pluripotency of human embryonic stem cells in the absence of feeder layers. *Biochem Biophys Res Commun* 330:934–942.

82. Wang, S. H., Tsai, M. S., Chiang, M. F., and Li, H. 2003. A novel NK-type homeobox gene, ENK early embryo specific NK., preferentially expressed in embryonic stem cells. *Gene Expr Patterns* 3:99–103.

83. Watcharasit, P., Bijur, G. N., Zmijewski, J. W., Song, L., Zmijewska, A., Chen, X., Johnson, G. V., and Jope, R. S. 2002. Direct, activating interaction between glycogen synthase kinase-3beta and p53 after DNA damage. *Proc Natl Acad Sci USA* 99:7951–7955.

84. Xiao, L., Yuan, X., and Sharkis, S. J. 2006. Activin A maintains self-renewal and regulates FGF, WnT and BMP pathways in human embryonic stem cells, *Stem Cells*.

85. Xu, R. H., Chen, X., Li, D. S., Li, R., Addicks, G. C., Glennon, C., Zwaka, T. P., and Thomson, J. A. 2002. BMP4 initiates human

embryonic stem cell differentiation to trophoblast. *Nat Biotechnol* 20:1261–1264.

86. Xu, R. H., Peck, R. M., Li, D. S., Feng, X., Ludwig, T., and Thomson, J. A. 2005. Basic FGF and suppression of BMP signaling sustain undifferentiated proliferation of human ES cells. *Nat Methods* 2:185–190.

87. Ying, Q. L., Nichols, J., Chambers, I., and Smith, A. 2003. BMP induction of Id proteins suppresses differentiation and sustains embryonic stem cell self-renewal in collaboration with STAT3. *Cell* 115:281–292.

88. Yu, J., Vodyanik, M. A., Smuga-Otto, K., Antosiewicz-Bourget, J., Frane, J. L., Tian, S., Nie, J., Jonsdottir, G. A., Ruotti, V., Stewart, R., Slukvin, II, and Thomson, J. A. 2007. Induced pluripotent stem cell lines derived from human somatic cells. *Science* 318:1917–1920.

89. Zhan, M., Miura, T., Xu, X., and Rao, M. S. 2005. Conservation and variation of gene regulation in embryonic stem cells assessed by comparative genomics. *Cell Biochem Biophys* 43:379–405.

90. Zhao, X. D., Han, X., Chew, J. L., Liu, J., Chiu, K. P., Choo, A., Orlov, Y. L., Sung, W. K., Shahab, A., Kuznetsov, V. A., Bourque, G., Oh, S., Ruan, Y., Ng, H. H., and Wei, C. L. 2007. Whole-genome mapping of histone H3 Lys4 and 27 trimethylations reveals distinct genomic compartments in human embryonic stem cells. *Cell Stem Cell* 1:286–298.

Chapter 5

Adult Stem/Progenitor Cell Self-Renewal

Jeffrey L. Spees, PhD

Department of Medicine and Stem Cell Core, University of Vermont, Colchester, VT 05446

For correspondence:
Dr. Jeffrey Spees
Department of Medicine
208 South Park Drive, Ste 2
Colchester, VT 05446

E-mail: Jeffrey.Spees@uvm.edu
Phone: (802) 656-2388
Fax: (802) 656-8932

Self-renewal is a key property of adult stem cells, allowing them to perpetuate themselves as undifferentiated cells for extended periods of time and also to replicate extensively in response to injury and disease. Stem cell self-renewal occurs through either asymmetric or symmetric cell divisions depending on the needs of the tissue and allows stem cells to generate progeny that possess similar self-renewal capacity and differentiation potential. Self-renewal can be regulated by cell-intrinsic mechanisms such as transcription factors, tumor suppressors, epigenetic modifiers, and microRNAs, or though cell-extrinsic mechanisms such as circulating factors, secreted paracrine factors, membrane bound ligands provided from neighboring cells, and extracellular matrix components. Some of the cell-extrinsic determinants of self-renewal are components of the stem cell niche, a specialized tissue structure that houses undifferentiated stem cells and that provides regulatory information to instruct quiescence, proliferation, mobilization, and differentiation [1].

Adult stem/progenitor cells differ from embryonic stem (ES) cells and induced pluripotent (iPS) cells both in their potency and in their degree of replicative potential. ES cells and iPS cells are pluripotent and capable of differentiating into any cell type of the body, while adult stem cells are multipotent and are capable of producing some or all of the cell types within their resident organ or tissue. Despite these differences, some of the mechanisms that confer stem cell properties to ES cells also operate in adult stem cells.

1. Sox2 Regulates Self-Renewal in ES Cells, iPS Cells, and Adult Stem Cells

ES cells self-renew and maintain themselves in an undifferentiated state in part through a network of proteins including Oct3/4 (POUF51), Sox2 (SRY [sex determining region Y]-box 2),

138

and *Nanog* (reviewed in [2]). These three transcription factors control the core transcriptional plan that confers ES cell properties [3]. *Oct3/4* is a critical determinant of self-renewal in ES cells [4] and must also be expressed to create iPS cells from adult somatic cells such as fibroblasts [5,6]. In contrast, genetic ablation studies suggest that the Oct3/4 protein is not necessary for self-renewal of adult stem cells [7]. *Nanog* affects pluripotency and the formation of germ cell derivatives from ES cells but is not required for self-renewal [8].

Sox2 is expressed by adult neural stem cells (NS cells) from both the lateral ventricle (subventricular zone) and hippocampal (subgranular zone) niches in the brain [9,10]. It is also expressed by adult astrocytes [11]. Reactive astrocytes exhibit NSC characteristics after brain injury [12]. Because they express *Sox2* as well as the cellular proliferation regulators *c-MYC* and *KLF4*, only *Oct3/4* expression is required to generate iPS cells from NS cells [13]. *Sox2* influences the proliferation of NS cells in part by interacting with the Notch and Sonic hedgehog (SHH) signaling pathways [14,15]. Other stem/progenitor cells of neuroectodermal origin also express *Sox2*, such as pituitary gland progenitor cells [16] and skin-derived precursor cells [17]. Adult endodermal derivatives that express *Sox2* include tracheal and bronchial progenitor (Clara) cells in the lung [18,19] and taste bud progenitor cells in the tongue [20,21]. Conditional deletion of *Sox2* leads to a reduced ability of basal stem cells in the trachea to proliferate after injury [18].

2. Epigenetic Regulation of Self-Renewal

Proteins from the polycomb and trithorax families are involved in chromatin remodeling and the regulation of gene expression. Polycomb members recruit repressor protein complexes and trithorax members recruit activating protein complexes that post-translationally modify histones (reviewed in [22]). The histone modifications affect the relative degree of chromatin condensation and accordingly the transcription of gene targets in compacted or uncompacted regions. The polycomb repressor *Bmi1* is highly expressed by both adult NS cells and hematopoietic stem cells (HS cells) and plays a key role in self-renewal (reviewed in [23]). *Bmi1* directly represses genes of the *Ink4a* locus such as *p16Ink4a* and *p19Arf* that are positive regulators of cell senescence [24]. *Bmi1* expression distinguishes primitive self-renewing NS cells and HS cells from downstream transit-amplifying progenitor cells. For example, *Bmi1* deficiency prevents the self-renewal of long-term repopulating HS cells, but allows for proliferation and survival of short-term HS cells that transiently reconstitute hematopoiesis in irradiated recipients that receive transplantation of fetal liver *Bmi1*$^{-/-}$ HS cells [25]. Similarly, *Bmi1* is required for self-renewal of NS cells but not for proliferation of restricted neural progenitors of the gut or forebrain [26]. In forebrain NS cells, *Bmi1* interacts with *Foxg1* to repress the *Cyclin-dependent kinase inhibitor 1A* (*p21, Cip1*), linking its effects to cell proliferation [27]. *Bmi1* may be a common regulator of adult stem/progenitor cells and is also expressed by pancreatic progenitor cells [28], lung progenitor cells [29], and intestinal stem cells [30]. Interestingly, *Bmi1* also alters mitochondrial function and the DNA damage response pathway, connecting the regulation of cell metabolism to self-renewal [31]. *Sall4* is a zinc finger transcription factor that regulates ES cell pluripotency by altering the expression of *Oct3/4* through its distal enhancer [32]. *Sall4* also activates the *Bmi1* promoter in adult HS cells, increasing their self-renewal [33].

The *Mixed Lineage Leukemia* (*Mll*) gene is a trithorax family member that was identified to be mutated in infant

140

leukemia [34]. Gene knockout of *Mll* in mice is embryonic lethal. Conditional deletion of *Mll* in adults leads to compromised self-renewal of HS cells upon transplantation into recipient animals, although hematopoiesis in the bone marrow, thymus, and spleen of the conditionally-deleted donors appears to be normal [35]. Therefore, similar to *Bmi1*, *Mll* may regulate self-renewal of primitive HS cells, but not the proliferation of downstream progenitor cells.

3. Signal Transduction Pathways Linked to Self-Renewal

The Notch, Wingless (WnT/beta-catenin), and SHH signal transduction pathways are thought to influence self-renewal in adult stem/progenitor cells from diverse tissues. However, assessing the role of each of these pathways in the self-renewal of particular adult stem cell systems has been controversial. Some of the controversy may be due to the use of different study models or to cellular responses that occur in culture but do not occur in vivo. In some cases, the conclusions reached from gain-of-function experiments do not match those of loss-of-function experiments. In other cases, a given signaling pathway may be active, but effects on stem cells may differ from effects on progenitor cells, sometimes leading to fate determination and differentiation in progenitors rather than self-renewal.

Notch signaling in mammals occurs through four Notch receptors (Notch 1–4) and five ligands (Jagged1–2 and Delta-like 1, 3, and 4) that interact with each other through cell-cell contact (reviewed in [36]). Self-renewal of NS cells is positively regulated by Notch signaling and subsequent receptor cleavage and release of the Notch intracellular domain (NICD) by an intramembrane protease complex called gamma secretase [37,38]. The gamma secretase complex includes: <u>presenilin</u> 1 or 2, <u>nicastrin</u>, anterior pharynx-defective 1 (<u>APH-1</u>), and presenilin enhancer 2 (PEN-2). Deletion of presenilin 1, the catalytic subunit of gamma secretase, was shown to significantly reduce NSC self-renewal [37]. Notch signaling affects both NS cells and downstream progenitor cells, but NS cells can be distinguished from progenitors by their increased sensitivity to the canonical Notch effector C-promoter binding factor 1 (CBF1) [39]. Notch signaling is also reported to increase the proliferation of skeletal muscle satellite cells (stem cells) and progenitor cells and to play a role in muscle repair [40,41].

Whether Notch signaling within the bone marrow niche is critical for HSC self-renewal is controversial [42]. Several initial reports with gain-of-function experiments suggested a key role for Notch signaling in both HSC self-renewal and in lineage commitment of hematopoietic progenitor cells. For example, HS cells transduced with virus in order to overexpress active Notch1 displayed a proliferative leukemic phenotype [43,44]. Enhanced Notch1 signaling was also observed to promote preferential lymphoid over myeloid lineage commitment from hematopoietic progenitor cells [44]. In further support for a role of Notch in HSC self-renewal, constituitive activation of the parathyroid receptor in osteoblasts, cells believed to contribute to the HSC niche, led to increased osteoblastic Jagged1 expression and increased HSC numbers [45]. Osteopontin (OPN) is a secreted matrix component of the HSC niche that was shown to negatively regulate HSC self-renewal. Deletion of OPN led to increased levels of Jagged1 in the niche and increased numbers of HS cells, again suggesting a role for Jagged1 in HSC proliferation [46]. In contrast, several loss-of-function experiments have painted a different picture for Notch and HS cells. Normal hematopoiesis was observed in mice with conditional deletions of numb and numb-like in HS cells [47]. These proteins regulate Notch receptor turnover and typically antagonize Notch signaling. Furthermore, expression of a dominant-negative Mastermind-like1, an inhibitor of Notch-mediated transcriptional activation in HS cells or, alternatively, removal of CSL/RBPJ, a DNA-binding factor that is required for canonical Notch signaling, did not alter HSC self-renewal or reconstitution by HSC progenitors [48].

WnT proteins are secreted and bind to the low-density lipoprotein receptors, LRP5 and LRP6, in addition to members of the Frizzled protein family (reviewed in [49]). In the absence of WnT signaling, the transcription factor beta-catenin is maintained in a complex with the scaffolding protein Axin and glycogen synthase kinase 3β (GSK3β) and phosphorylated. Phosphorylated beta-catenin is ubiquitinated and subsequently degraded by the proteosome. In the presence of WnT ligand, GSK3β does not phosporylate beta-catenin, and it instead accumulates in a free state, localizes to the nucleus, and induces gene transcription by binding to LEF/TCF transcription factors. WnT signaling is important for self-renewal of multiple adult stem cells including NS cells [50], HS cells [51,52], and intestinal stem cells [53]. For lung bronchiolar progenitor cells, gain-of-function experiments with a constituitively active form of beta-catenin suggested a role for WnT signaling in lung epithelial repair [54]. However, subsequent loss-of-function studies determined that the bronchiolar epithelium could maintain itself and repair itself without the need for beta-catenin signaling [55]. SHH signaling regulates NSC self-renewal [15] but was found to be dispensable for HSC self-renewal [56].

4. MicroRNAs and Adult Stem Cell Self-Renewal

MicroRNAs (miRNAs) have emerged as a general regulatory system that fine-tunes protein expression in cells. miRNAs are small 22 or 23 nucleotide RNAs that downregulate protein expression post-transcriptionally through binding to mRNA sequences (reviewed in [57,58]). In ES cells and adult stem cells, miRNAs affect both differentiation and self-renewal (reviewed in [59]). In ES cells, miRNAs promoters have been mapped to the genome regions for the core regulatory proteins that govern pluripotency [60]. By examining miRNA levels that increased during ES cell differentiation, mi-145 was identified as a regulator of *Oct3/4*, *Sox2*, and *KLF4*. Furthermore, loss of mi-145 impaired differentiation, and overexpression of mi-145 was demonstrated to reduce ES cell self-renewal [61]. In mammary stem cells, miRNA-200c was found to regulate the expression of *Bmi1* and to suppress the ability of the cells to form mammary ducts [62]. The orphan nuclear receptor TLX was found to increase NSC self-renewal, in part though WnT/beta-catenin signaling [63]. TLX expression was suppressed by miRNA-9, and miRNA-9 localized to neurogenic areas of the brain. In a feedback loop, miRNA-9 levels were demonstrated to be repressed by TLX [64]. Compared with levels found in fetal and young mice, the DNA binding protein Hmga2 was found to be reduced in concentration in adults, leading to increased levels of *p16Ink4a* and *p19Arf* and decreased proliferation of NS cells. *Hmga2* levels were shown to be repressed by the let 7b miRNA, a miRNA with expression that increases with age [65]. These studies reveal an interesting link between miRNAs, adult stem cell self-renewal, and ageing.

References

1. Scadden D. T. 2006. The stem-cell niche as an entity of action. *Nature* 441(7097):1075–9. Review.

2. Shenghui H., Nakada D., and Morrison S. J. 2009. Mechanisms of stem cell self-renewal. *Annu Rev Cell Dev Biol* 25:377–406. Review.

3. Boyer L. A., Lee T. I., Cole M. F., Johnstone S. E., Levine S. S., Zucker J. P., Guenther M. G., Kumar R. M., Murray H. L., Jenner R. G., Gifford D. K., Melton D. A., Jaenisch R., and Young R. A. 2005. Core transcriptional regulatory circuitry in human embryonic stem cells. *Cell* 122(6):947–56.

4. Niwa H., Miyazaki J., and Smith A. G. 2000. Quantitative expression of Oct-3/4 defines differentiation, dedifferentiation or self-renewal of ES cells. *Nat Genet* 24(4):372–6.

5. Takahashi K. and Yamanaka S. 2006. Induction of pluripotent stem cells from mouse embryonic and adult fibroblast cultures by defined factors. *Cell* 126(4):663–76. Epub Aug 10.

6. Takahashi K., Tanabe K., Ohnuki M., Narita M., Ichisaka T., Tomoda K., and Yamanaka S. 2007. Induction of pluripotent stem cells from adult human fibroblasts by defined factors. *Cell* 131(5):861–72.

7. Lengner C. J., Camargo F. D., Hochedlinger K., Welstead G. G., Zaidi S., Gokhale S., Scholer H. R., Tomilin A., and Jaenisch R. 2007. *Oct4* expression is not required for mouse somatic stem cell self-renewal. *Cell Stem Cell* 1(4):403–15.

8. Chambers I., Silva J., Colby D., Nichols J., Nijmeijer B., Robertson M., Vrana J., Jones K., Grotewold L., and Smith A. 2007. Nanog safeguards pluripotency and mediates germline development. *Nature* 450(7173):1230–4.

9. Graham V., Khudyakov J., Ellis P., and Pevny L. 2003. SOX2 functions to maintain neural progenitor identity. *Neuron* 39(5):749–65.

10. Suh H., Consiglio A., Ray J., Sawai T., D'Amour K. A., and Gage F. H. 2007. In vivo fate analysis reveals the multipotent and self-renewal capacities of Sox2+ neural stem cells in the adult hippocampus. *Cell Stem Cell* 1(5):515–28.

11. Komitova M. and Eriksson P. S. 2004. Sox-2 is expressed by neural progenitors and astroglia in the adult rat brain. *Neurosci Lett* 369(1):24–7.

12. Buffo A., Rite I., Tripathi P., Lepier A., Colak D., Horn A. P., Mori T., and Götz M. 2008.Origin and progeny of reactive gliosis: A source of multipotent cells in the injured brain. *Proc Natl Acad Sci USA* 105(9):3581–6. Epub 2008 Feb 25.

13. Kim J. B., Sebastiano V., Wu G., Araúzo-Bravo M. J., Sasse P., Gentile L., Ko K., Ruau D., Ehrich M., van den Boom D., Meyer J., Hübner K., Bernemann C., Ortmeier C., Zenke M., Fleischmann B. K., Zaehres H., and Schöler H. R. 2009. *Oct4*-induced pluripotency in adult neural stem cells. *Cell* 136(3):411–9.

14. Bani-Yaghoub M., Tremblay R. G., Lei J. X., Zhang D., Zurakowski B., Sandhu J. K., Smith B., Ribecco-Lutkiewicz M., Kennedy J., Walker P. R., and Sikorska M. 2006. Role of Sox2 in the development of the mouse neocortex. *Dev Biol* 295(1):52–66. Epub 2006 Apr 24.

15. Favaro R., Valotta M., Ferri A. L., Latorre E., Mariani J., Giachino C., Lancini C., Tosetti V., Ottolenghi S., Taylor V., and Nicolis S. K. 2009. Hippocampal development and neural stem cell maintenance require Sox2-dependent regulation of Shh. *Nat Neurosci* 12(10):1248–56. Epub 2009 Sep 6.

16. Fauquier T., Rizzoti K., Dattani M., Lovell-Badge R., and Robinson I. C. 2008. SOX2-expressing progenitor cells generate all of the major cell types in the adult mouse pituitary gland. *Proc Natl Acad Sci USA* 105(8):2907–12. Epub 2008 Feb 15.

17. Biernaskie J., Paris M., Morozova O., Fagan B. M., Marra M., Pevny L., and Miller F. D. 2009. SKPs derive from hair follicle precursors and exhibit properties of adult dermal stem cells. *Cell Stem Cell* 5(6):610–23.

18. Que J., Luo X., Schwartz R. J., and Hogan B. L. 2009. Multiple roles for Sox2 in the developing and adult mouse trachea. *Development* 136(11):1899–907. Epub 2009 Apr 29.

19. Tompkins D. H., Besnard V., Lange A. W., Wert S. E., Keiser A. R., Smith A. N., Lang R., and Whitsett J. A. Sox2 is required for maintenance and differentiation of bronchiolar Clara, ciliated, and goblet cells. PLoS One 4(12):e8248.

20. Okubo T., Clark C., and Hogan B. L. 2009. Cell lineage mapping of taste bud cells and keratinocytes in the mouse tongue and soft palate. *Stem Cells* 27(2):442–50.

21. Okubo T., Pevny L. H., and Hogan B. L. 2006. Sox2 is required for development of taste bud sensory cells. *Genes Dev* 20(19):2654–9.

22. Schuettengruber B., Chourrout D., Vervoort M., Leblanc B., and Cavalli G. 2007. Genome regulation by polycomb and trithorax proteins. *Cell* 128(4):735–45. Review.

23. Park I. K., Morrison S. J., Clarke M. F. 2004. Bmi1, stem cells, and senescence regulation. *J Clin Invest* 113(2):175–9. Review.

24. Molofsky A. V., He S., Bydon M., Morrison S. J., and Pardal R. 2005. Bmi-1 promotes neural stem cell self-renewal and neural development but not mouse growth and survival by repressing the p16Ink4a and p19Arf senescence pathways. *Genes Dev* 19(12):1432–7.

25. Park I. K., Qian D., Kiel M., Becker M. W., Pihalja M., Weissman I. L., Morrison S. J., and Clarke M. F. 2003. Bmi-1 is required for maintenance of adult self-renewing haematopoietic stem cells. *Nature* 423(6937):302–5. Epub 2003 Apr 20.

26. Molofsky A. V., Pardal R., Iwashita T., Park I. K., Clarke M. F., and Morrison S. J. 2003. Bmi-1 dependence distinguishes neural stem cell self-renewal from progenitor proliferation. *Nature* 425(6961):962–7. Epub 2003 Oct 22.

27. Fasano C. A., Phoenix T. N., Kokovay E., Lowry N., Elkabetz Y., Dimos J. T., Lemischka I. R., Studer L., and Temple S. 2009. Bmi-1 cooperates with Foxg1 to maintain neural stem cell self-renewal in the forebrain. *Genes Dev* 23(5):561–74.

28. Sangiorgi E. and Capecchi M. R. 2009. Bmi1 lineage tracing identifies a self-renewing pancreatic acinar cell subpopulation capable of maintaining pancreatic organ homeostasis. *Proc Natl Acad Sci USA* 106(17):7101–6. Epub 2009 Apr 16.

29. Dovey J. S., Zacharek S. J., Kim C. F., and Lees J. A. 2008. Bmi1 is critical for lung tumorigenesis and bronchioalveolar stem cell expansion. *Proc Natl Acad Sci USA* 105(33):11857–62. Epub 2008 Aug 12.

30. Sangiorgi E. and Capecchi M. R. 2008. Bmi1 is expressed in vivo in intestinal stem cells. *Nat Genet* 40(7):915–20. Epub 2008 Jun 8.

31. Liu J., Cao L., Chen J., Song S., Lee I. H., Quijano C., Liu H., Keyvanfar K., Chen H., Cao L. Y., Ahn B. H., Kumar N. G., Rovira I. I., Xu X. L., Van Lohuizen M., Motoyama N., Deng C. X., and Finkel T. 2009. Bmi1 regulates mitochondrial function and the DNA damage response pathway. *Nature* 459(7245):387–92. Epub 2009 Apr 29.

32. Zhang J., Tam W. L., Tong G. Q., Wu Q., Chan H. Y., Soh B. S., Lou Y., Yang J., Ma Y., Chai L., Ng H. H., Lufkin T., Robson P., and Lim B. 2006. Sall4 modulates embryonic stem cell pluripotency and early embryonic development by the transcriptional regulation of Pou5f1. *Nat Cell Biol* 8(10):1114–23. Epub 2006 Sep 17.

33. Yang J., Chai L., Liu F., Fink L. M., Lin P., Silberstein L. E., Amin H. M., Ward D. C., and Ma Y. 2007. Bmi-1 is a target gene for SALL4 in hematopoietic and leukemic cells. *Proc Natl Acad Sci USA* 104(25):10494–9. Epub 2007 Jun 8.

34. Tkachuk D. C., Kohler S., and Cleary M. L. 1992. Involvement of a homolog of Drosophila trithorax by 11q23 chromosomal translocations in acute leukemias. *Cell* 71(4):691–700.

35. McMahon K. A., Hiew S. Y., Hadjur S., Veiga-Fernandes H., Menzel U., Price A. J., Kioussis D., Williams O., and Brady H. J. 2007. Mll has a critical role in fetal and adult hematopoietic stem cell self-renewal. *Cell Stem Cell* 1(3):338–45.

36. Baron M. 2003. An overview of the Notch signalling pathway. *Semin Cell Dev Biol* 14(2):113–9. Review.

37. Hitoshi S., Alexson T., Tropepe V., Donoviel D., Elia A. J., Nye J. S., Conlon R. A., Mak T. W., Bernstein A., and van der Kooy D. 2002. Notch pathway molecules are essential for the maintenance, but not the generation, of mammalian neural stem cells. *Genes Dev* 16(7):846–58.

38. Androutsellis-Theotokis A., Leker R. R., Soldner F., Hoeppner D. J., Ravin R., Poser S. W., Rueger M. A., Bae S. K., Kittappa R., and McKay R. D. 2006. Notch signalling regulates stem cell numbers in vitro and in vivo. *Nature* 442(7104):823–6. Epub 2006 Jun 25.

39. Mizutani K., Yoon K., Dang L., Tokunaga A., and Gaiano N. 2007. Differential Notch signalling distinguishes neural stem cells from intermediate progenitors. *Nature* 449(7160):351–5. Epub 2007 Aug 26.

40. Conboy I. M. and Rando T. A. 2002. The regulation of Notch signaling controls satellite cell activation and cell fate determination in postnatal myogenesis. *Dev Cell* 3(3):397–409.

41. Conboy I. M., Conboy M. J., Wagers A. J., Girma E. R., Weissman I. L., and Rando T. A. 2005. Rejuvenation of aged progenitor cells by exposure to a young systemic environment. *Nature* 433(7027):760–4.

42. Weber J. M. and Calvi L. M. 2009. Notch signaling and the bone marrow hematopoietic stem cell niche. *Bone* Aug 11. [Epub ahead of print]

43. Varnum-Finney B., Xu L., Brashem-Stein C., Nourigat C., Flowers D., Bakkour S., Pear W. S., and Bernstein I. D. 2000. Pluripotent, cytokine-dependent, hematopoietic stem cells are immortalized by constitutive Notch1 signaling. *Nat Med* 6(11):1278–81.

44. Stier S., Cheng T., Dombkowski D., Carlesso N., and Scadden D. T. 2002. Notch1 activation increases hematopoietic stem cell self-renewal in vivo and favors lymphoid over myeloid lineage outcome. *Blood* 99(7):2369–78.

45. Calvi L. M., Adams G. B., Weibrecht K. W., Weber J. M., Olson D. P., Knight M. C., Martin R. P., Schipani E., Divieti P., Bringhurst F. R., Milner L. A., Kronenberg H. M., and Scadden D. T. 2003. Osteoblastic cells regulate the haematopoietic stem cell niche. *Nature* 425(6960):841–6.

46. Stier S., Ko Y., Forkert R., Lutz C., Neuhaus T., Grünewald E., Cheng T., Dombkowski D., Calvi L. M., Rittling S. R., and Scadden D. T. 2005. Osteopontin is a hematopoietic stem cell niche component that negatively regulates stem cell pool size. *J Exp Med* 201(11):1781–91. Epub 2005 May 31.

47. Wilson A., Ardiet D. L., Saner C., Vilain N., Beermann F., Aguet M., Macdonald H. R., and Zilian O. 2007. Normal hemopoiesis and lymphopoiesis in the combined absence of numb and numblike. *J Immunol* 178(11):6746–51.

48. Maillard I., Koch U., Dumortier A., Shestova O., Xu L., Sai H., Pross S. E., Aster J. C., Bhandoola A., Radtke F., and Pear W. S. 2008. Canonical notch signaling is dispensable for the maintenance of adult hematopoietic stem cells. *Cell Stem Cell* 2(4):356–66.

49. Clevers H. 2006. WnT/beta-catenin signaling in development and disease. *Cell* 127(3):469–80. Review.

50. Kalani M. Y., Cheshier S. H., Cord B. J., Bababeygy S. R., Vogel H., Weissman I. L., Palmer T. D., and Nusse R. 2008. WnT-mediated self-renewal of neural stem/progenitor cells. *Proc Natl Acad Sci USA* 105(44):16970–5. Epub 2008 Oct 28.

51. Reya T., Duncan A. W., Ailles L., Domen J., Scherer D. C., Willert K., Hintz L., Nusse R., and Weissman I. L. 2003. A role for WnT signalling in self-renewal of haematopoietic stem cells. *Nature* 423(6938):409–14. Epub 2003 Apr 27.

52. Fleming H. E., Janzen V., Lo Celso C., Guo J., Leahy K. M., Kronenberg H. M., and Scadden D. T. 2008. WnT signaling in the niche enforces hematopoietic stem cell quiescence and is necessary to preserve self-renewal in vivo. *Cell Stem Cell* 2(3):274–83.

53. Van der Flier L. G., Van Gijn M. E., Hatzis P., Kujala P., Haegebarth A., Stange D. E., Begthel H., Van den Born M., Guryev V., Oving I., Van Es J. H., Barker N., Peters P. J., Van de Wetering M., and Clevers H. 2009. Transcription factor achaete scute-like 2 controls intestinal stem cell fate. *Cell* 136(5):903–12.

54. Reynolds S. D., Zemke A. C., Giangreco A., Brockway B. L., Teisanu R. M., Drake J. A., Mariani T., Di P. Y., Taketo M. M., and Stripp B. R. 2008. Conditional stabilization of beta–

catenin expands the pool of lung stem cells. *Stem Cells* 26(5):1337–46. Epub 2008 Mar 20.

55. Zemke A. C., Teisanu R. M., Giangreco A., Drake J. A., Brockway B. L., Reynolds S. D., and Stripp B. R. 2009. beta-Catenin is not necessary for maintenance or repair of the bronchiolar epithelium. *Am J Respir Cell Mol Biol* 41(5):535–43. Epub 2009 Feb 12.

56. Gao J., Graves S., Koch U., Liu S., Jankovic V., Buonamici S., El Andaloussi A., Nimer S. D., Kee B. L., Taichman R., Radtke F., and Aifantis I. 2009. Hedgehog signaling is dispensable for adult hematopoietic stem cell function. *Cell Stem Cell* 4(6):548–58.

57. Bartel D. P. 2004. MicroRNAs: genomics, biogenesis, mechanism, and function. *Cell* 116(2):281–97. Review.

58. Bartel D. P. 2009. MicroRNAs: target recognition and regulatory functions. *Cell* 136(2):215–33. Review.

59. Gangaraju V. K. and Lin H. 2009. MicroRNAs: key regulators of stem cells. *Nat Rev Mol Cell Biol* 10(2):116–25. Review.

60. Marson A., Levine S. S., Cole M. F., Frampton G. M., Brambrink T., Johnstone S., Guenther M. G., Johnston W. K., Wernig M., Newman J., Calabrese J. M., Dennis L. M., Volkert T. L., Gupta S., Love J., Hannett N., Sharp P. A., Bartel D. P., Jaenisch R., and Young R. A. 2008. Connecting microRNA genes to the core transcriptional regulatory circuitry of embryonic stem cells. *Cell* 134(3):521–33.

61. Xu N., Papagiannakopoulos T., Pan G., Thomson J. A., and Kosik K. S. MicroRNA-145 regulates *OCT4*, *SOX2*, and *KLF4* and represses pluripotency in human embryonic stem cells. Cell. 2009 May 15;137(4):647–58. Epub 2009 Apr 30.

62. Shimono Y., Zabala M., Cho R. W., Lobo N., Dalerba P., Qian D., Diehn M., Liu H., Panula S. P., Chiao E., Dirbas F. M., Somlo G., Pera R. A., Lao K., and Clarke M. F. 2009. Downregulation of miRNA-200c links breast cancer stem cells with normal stem cells. *Cell* 138(3):592–603.

63. Qu Q., Sun G., Li W., Yang S., Ye P., Zhao C., Yu R. T., Gage F. H., Evans R. M., and Shi Y. 2010. Orphan nuclear recep-

tor TLX activates WnT/beta-catenin signalling to stimulate neural stem cell proliferation and self-renewal. *Nat Cell Biol* 12(1):31–40; sup pp 1–9. Epub 2009 Dec 13.

64. Zhao C., Sun G., Li S., and Shi Y. 2009. A feedback regulatory loop involving microRNA-9 and nuclear receptor TLX in neural stem cell fate determination. *Nat Struct Mol Biol* 16(4):365–71. Epub 2009 Mar 29.

65. Nishino J., Kim I., Chada K., and Morrison S. J. 2008. Hmga2 promotes neural stem cell self-renewal in young but not old mice by reducing p16Ink4a and p19Arf Expression. *Cell* 135(2):227–39.

Chapter 6

Embryonic Stem Cells, Nuclear Cloning, and Induced Pluripotent Stem Cells

Miguel Angel Esteban[1] and Zhong Wang[2]

[1]*Stem Cell and Cancer Biology Group, Key Laboratory of Regenerative Biology, South China Institute for Stem Cell Biology and Regenerative Medicine, Guangzhou Institutes of Biomedicine and Health, Chinese Academy of Sciences, Guangzhou 510530, China,* maestebanb@gmail.com (email), +86-20-3205-1820 (phone), 86-20-3201-5299 (Fax) ;

[2]Cardiovascular Research Center, Massachusetts General Hospital, Harvard Medical School and Harvard Stem Cell Institute, Richard Simches Research Center, 185 Cambridge Street, Boston, MA 02114. zhwang@partners.org (email), 617-643-3444 (phone), 616-643-3471 (fax).

1. Origin and Property of Embryonic Stem Cells

Embryonic stem (ES) cells are pluripotent cell lines derived from early stage mammalian embryos [1]. During development, the fertilized metazoan zygote undergoes a series of symmetric cell divisions and compaction (morula stage) to soon go through the first differentiation process: the separation into the inner cell mass (ICM) and the trophectoderm [2] (Figure 1. Early embryonic development). The ICM will produce the fetus, and the trophoectoderm will form the placenta. Afterwards, the ICM differentiates into a lower and upper layer, the hypoblast and epiblast, respectively. The hypoblast will form the yolk sac, the umbilical cord, and the blood vessels of the placenta. Part of the epiblast will generate the amnios (the cavity that contains the embryo) while the remaining cells will transform into the so-called three embryonic layers (ectoderm, mesoderm, and endoderm) through a process known as gastrulation. In humans, gastrulation occurs early after implantation into the maternal decidua (around day 14–16) (figure 1). The ectoderm is the first to emerge among all three layers and it forms the epidermis, the nervous system, and the skeleton of the head. The endoderm produces the epithelial cells of all organs and glands, and the mesoderm generates most bones and muscles, the blood and the circulatory system, and all connective tissues [2].

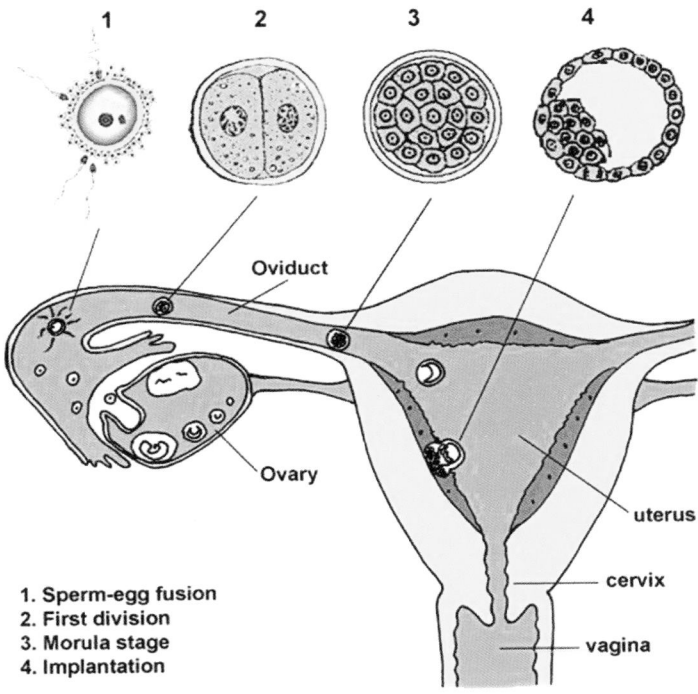

Figure 1: Schematic representation of early embryonic development from the moment of fertilization. The morula stage and the formation of the ICM are also depicted. Pluripotent stem cells from the ICM will later on form all tissues that form an adult body.

ES cell lines are derived from the ICM of the blastocyst stage embryo and are considered their *in vitro* counterparts [1]. ES cells have an enticing therapeutic potential in regenerative medicine because, like ICM cells, they are pluripotent—they possess the ability to differentiate into all the cell types that compose an adult body (figure 1). Another unique feature of ES cells is their ability to divide indefinitely and remain uncommitted to any cell types (self-renewal), if cultured under the right conditions, thus providing unlimited cell sources for potential stem cell–based therapies or *in vitro* disease modeling

[3]. The first ES cell lines were derived from mice in 1981 [4], and their human counterparts (hES cells) were established in 1998 from *in vitro* fertilized eggs originally kept for reproductive purposes [5]. ES cells from monkeys and more recently rats have also been isolated [6, 7], but this has not yet been achieved with other important species for biomedical research, like swine. Because the ICM stage is very transient *in vivo*, a key challenge in isolating mouse and human ES cell lines has been to find appropriate conditions that allow indefinite proliferation and self-renewal *in vitro*. Interestingly, mouse ES cells can also be derived from the epiblast using culture conditions more similar to hES cells [8]. Notably, hES cells are more similar to mouse epiblast ES cells than to mouse ES cells, which has raised the question as to whether hES cells are truly ES cells. Such consideration is important because even though mouse epiblast ES cells can form teratomas (tumors composed of tissues derived from the three germ layers) when injected into immune-depressed mice, they cannot contribute to tissue formation after injection into heterologous blastocysts (Figure 2. Teratomas derived from mouse ES cells)

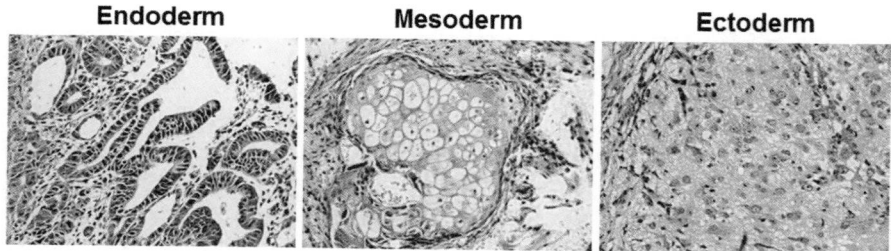

Figure 2: Mouse ES cells (R1 cell line) were injected subcutaneously into immune-compromised mice and teratomas developed after 3 weeks. Sections were stained with hematoxylin-eosin and show derivatives of the 3 germ layers (gland-like structures, cartilage, and neural-like cells, in this order).

156

2. Patient-Specific Pluripotent Stem Cells

The use of hES cells in regenerative medicine has faced several major obstacles [9]. One major challenge is the immune-compatibility between the acceptor and the donor hES cell line to prevent rejection (Figure 3. Patient-specific ES cells). Making this feasible would imply the need for a huge bank of hES cells corresponding to all possible immunological variants. This is impractical for technical reasons and could also receive fierce opposition given the associated ethical considerations. Another major challenge is the bona fide differentiation of ES cells into specific lineages (e.g., neurons or pancreatic beta cells) that can integrate and function properly after transplantation in patients. Nowadays, existing protocols for tissue specific–differentiation are evolving but they are in most cases both time-consuming and inefficient. For transplanting differentiated cells, aside from the immune-compatibility issue, high risk of teratoma formation is another concern unless the newly differentiated cells have been filtered adequately and are stable [10]. Remarkably, differentiated retina cells from hES cells were recently succesfully transplanted into affected individuals, with apparent stability and adequate function [11].

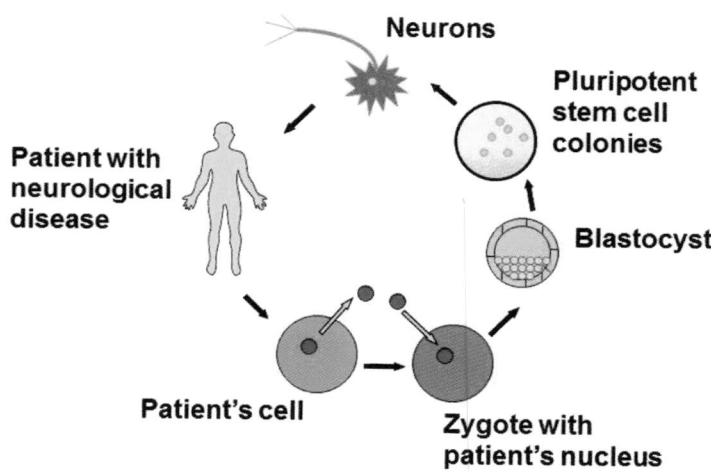

Figure 3: Generation of patient specific ES cells (in the scheme represented as if it was achieved by means of nuclear transfer) would allow autologous transplantation of *in vitro* generated cell lineages devoid of ethical considerations and immune rejection.

Despite the difficulties of transplantation, hES cell lines generated from blastocysts bearing mutations that cause human disease could represent an outstanding model to study the diseased state *in vitro* [12] (figure 3). This is particularly relevant for those diseases that lack animal models. For example, neurons differentiated from patient-specific ES cells could potentially mimic some aspects of the disease *in vitro*, and this could offer the chance to understand the underlying mechanisms and provide cell sources for high-throughput screening of compounds to correct cell function. However, most hES cell lines created thus far do not carry patient-specific diseases. Therefore, a major scientific goal of recent years has been to generate individualized patient-specific hES cells from somatic cells that can be used for either autologous transplantation or to study human diseases. Several methods of nuclear reprogramming (i.e., reprogramming of a somatic cell to an embryonic-like stage) have been proposed and explored in the mouse and human systems, among them nuclear cloning (also called somatic cell nuclear transfer, or SCNT) and induced pluripotent stem cells (iPSCs) are the most extensively studied and have generated great interest.

3. Nuclear Cloning

Nuclear cloning/SCNT (Figure 4. Nuclear cloning) was the first choice for generating patient-specific hES cells because of its long history and the successful cloning of Dolly in 1997 [13]. Several decades earlier, pioneer studies by Briggs, King, and Gurdon had showed that by transferring a somatic nucleus to a frog egg deprived of its own nucleus, the egg bearing the somatic nuclear DNA was able to develop into a heartbeat-stage tadpole [14]. The birth of Dolly the sheep was the first demonstration that SCNT can generate an entire individual from an adult somatic cell. Altogether, these studies revealed that somatic cells contain the same genetic information as those in fertilized eggs to direct the development of the whole organism. During SCNT, the somatic nucleus is reset, or reprogrammed, back to its embryonic stage. This reprogramming does not involve any changes in genetic material and thus has been called epigenetic reprogramming, which will be discussed in more detail later in this chapter.

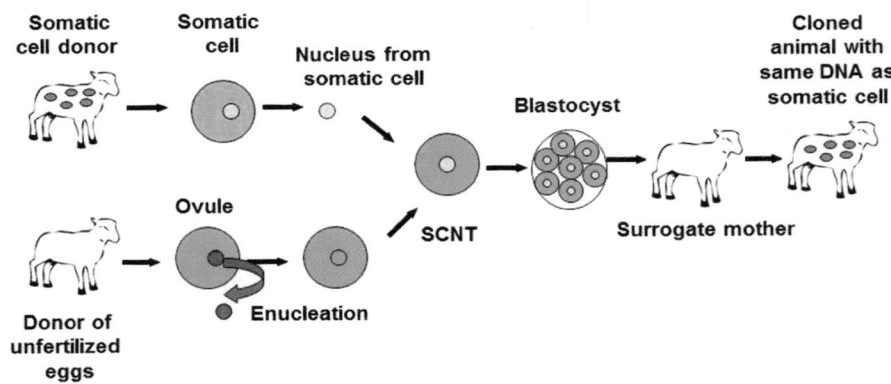

Figure 4: Simplified representation of the nuclear cloning procedure used for producing Dolly the sheep in 1997.

Since embryos generated by SCNT can undergo mitotic cleavage and become blastocyst under culture, this technology has been explored to generate ES cells in many species. In mice, although derived animals show certain health problems, the corresponding cell lines are indistinguishable from ES cells obtained from normal fertilized embryos [15]. For example, they can contribute to every tissue judged by chimera assays and even form an entire embryo after tetraploid complementation experiments (the technique is explained below) (Figure 5. Chimera and tetraploid complementation assays).

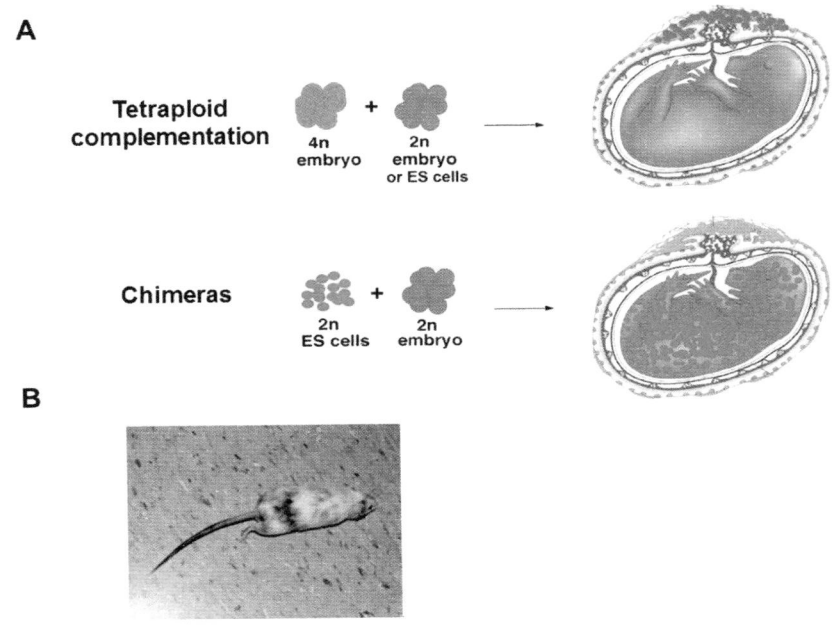

Figure 5 : A, Embryo-embryo or ES cell-embryo aggregation experiments.
When a tetraploid (4n) embryo are aggregated with a diploid embryos or
diploid ES cells, cells from the tetraploid embryos generally only contribute to placenta. On the other hand, if diploid ES cells are aggregated with
a diploid embryo and put back into uterus, ES cells will normally contribute
to the embryo proper during later development. Tetraploid complementation assay is the most stringent test for identifying bona fide mouse ES
and induced pluripotent cells. B, Captures of a chimeric mice produced
by injection of iPSC (black mouse) into the blastocysts of a white mouse.

Despite the scarcity of material and ethical concerns, nuclear cloning was vigorously applied in humans in an attempt to generate patient-specific hES cells. Unfortunately,
even though the procedure has been successfully achieved
in large animals, including pigs and nonhuman primates, it
appears technically more challenging in humans, and early
reports turned out to be fraudulent. Nevertheless, it is con-
ceivable that eventually human ES cells can be created
successfully through SCNT, and in fact this has been already
achieved if the oocyte nucleus is left unremoved [16].

161

4. Induced Pluripotent Stem Cells

An unexpected and ground-breaking study led by Shinya Yamanaka in 2006 discovered that somatic cells can be directly reprogrammed into ES cell–like cells by defined transcription factors [17]. Nuclear transfer had demonstrated that factors present in the freshly fertilized eggs can reprogram the nucleus of a somatic cell back to its embryonic stage. This immediately created enormous interest in the molecular mechanisms that govern ES cell behavior. As a result, a series of transcription factors including Oct4, Nanog, and Sox2 were identified to be essential for ES cell function [18]. Other important ES cell regulators have been discovered subsequently, including microRNAs and chromatin modifying factors (Figure 6. Molecular mechanisms of ES cell pluripotency). These factors/molecules together orchestrate a complicated network full of crosstalk.

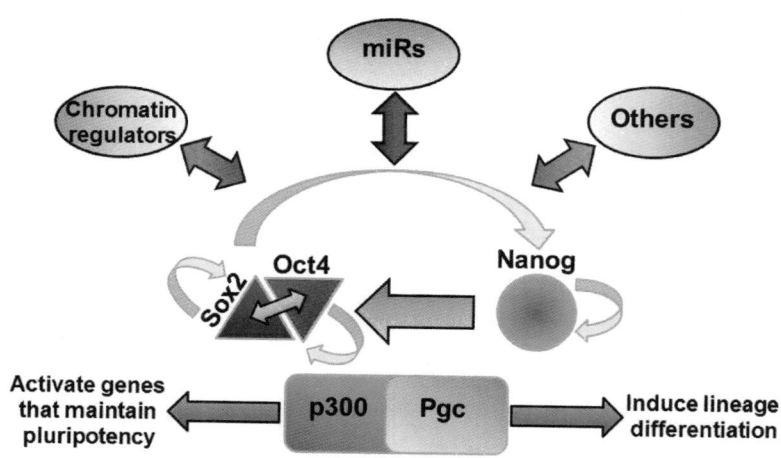

Figure 6: ES cell pluripotency is regulated by a complicated network of transcription factors (i.e. Sox2, Oct4, and Nanog), chromatin modifiers, microRNAs (miRs; i.e. miR302 family) and other proteins. ES cell specific transcription factors can be activators or repressors depending on the gene context. In the case of gene activation they normally do this by interacting with the

protein p300, while their interaction with polycomb group complex (Pgc) genes promotes gene repression.

Knowledge of the transcription network guiding ES cell pluripotency inspired Shinya Yamanaka to reason that somatic cells might be directly reprogrammed into pluripotent ES cell–like cells by over-expressing a defined group of transcription factors. Yamanaka and collaborators first transduced mouse embryonic fibroblasts with a cocktail of retroviruses individually expressing twenty-four transcription factors and proteins highly enriched in ES cells. Infected fibroblasts were then cultured under mouse ES cell culture conditions until ES cell–like colonies appeared, and these colonies were then later picked and expanded [17]. Sequential elimination of factors demonstrated that the combination of *Oct4*, *Sox2*, *Klf4*, and c-Myc was optimal for producing these cell lines (termed induced pluripotent stem cells, or iPS cells) (Figure 7. Induced pluripotent stem cells).

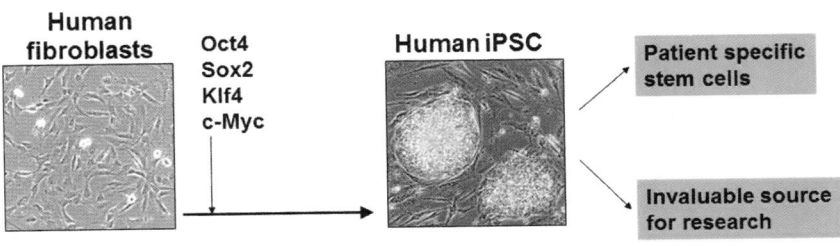

Figure 7 : Generation of human iPSC using adult skin fibroblasts and the 4 Yamanaka transcription factors delivered by means of retroviral vectors. hES cell-like colonies appeared at day 18 and were picked for further expansion at day 25 (in the picture)

The first generation iPS cells mimicked some aspects of mouse ES cells, including teratoma formation, but were not chimera competent. The technique was then improved by several laboratories, and the resulting mouse iPS cells were chimera competent and had germ line transmission [19, 20]. The transcriptome of iPS cells was very similar to mouse ES cells, and the expression of the exogenous factors was silenced by DNA methylation, a self-defense mechanism of ES cells to viral invading genomes. Subsequent studies have demonstrated that mouse iPS cells can produce entire animals in mouse tetraploid complementation experiments [21]. In such experiments, the cells from the tetraploid embryo will only contribute to the placenta, whereas the whole embryo proper can be derived entirely from the injected ES cells or iPS cells (figure 5). Ever since, iPS cells from multiple species including human, monkey, rat, and pigs have been generated using multiple methods (involving DNA integration or not) and tissue sources [22, 23]. Moreover, iPS cells generated from patients with genetic diseases have reproduced aspects of the disease *in vitro* and the number of disease-specific iPS cells generated from patients is growing steadily [24, 25].

Although producing chimera competent mouse iPS cells is not difficult, generating human iPS cells has turned out to be significantly more inefficient [26]. The low efficiency combined with the lack of standards for defining full reprogramming in the human context has also raised many concerns. Several studies have shown that inhibition of the p53 pathway can significantly improve iPS cell formation, but this may further increase the risk of tumorigenesis due to incorporation of mutations [27]. In this regard, and possibly related to the proliferation stress that the reprogramming involves, it has been reported that human iPS cells contain somatic point mutations, copy number variations and epigenetic alterations [28]. However, it remains to be studied whether these abnormalities are general or can be circumvented using specific donor cell sources and reprogramming methods.

For all these reasons, it is clear that extreme precautions and extensive animal trials will be required before iPS cells can be used in human clinical trials. For this purpose, preclinical studies with iPS cells derived from species like swine and nonhuman primates—whose physiology is comparable to humans—may be very important, as these cells resemble hES cells [22] (Figure 8. Utility of porcine iPS cells). As mentioned above, pig ES cells have not yet been succesfully isolated [29].

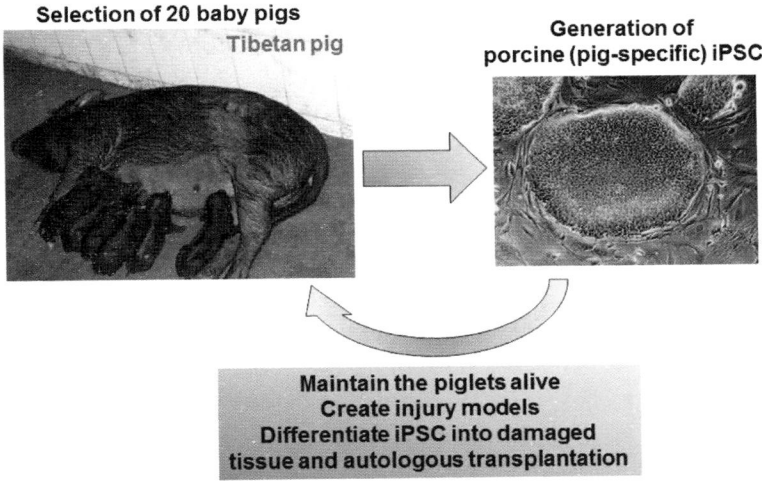

Figure 8: Use of pig iPS cells may provide the bridge between mouse and human iPS cells. Given their small size and different physiology mouse is not a good model for preclinical trials before iPS cell is potentially tested in human. Use of a large animal like the pig may thus provide invaluable.

5. ES Cells, iPS Cells, and Epigenetic Landscape

Regardless of the cell source, either ES cells or iPS cells, a fundamental barrier for understanding their limitations and realizing their potential is our limited knowledge of the molecular mechanisms determining pluripotency or differentiation into tissue-specific cell types. In this regard, exploring the epigenetic events that accompany somatic cell reprogramming may help us understand the basic developmental biology and find better ways for efficient directed differentiation.

SCNT and iPS cell generation can be viewed under the concept of "epigenetic landscape," originally proposed by Conrad Waddington in 1957 [30] (Figure 9. Waddington's epigenetic landscape). The epigenetic landscape was used to describe cell differentiation decisions during development even before the physical structure of genes had been defined. Under this view, the development of a fertilized egg or ES/iPS cell into a particular mature cell type is like a ball rolling through a defined trajectory path from the top of a hill into the bottom of a valley. Conversely, nuclear reprogramming can be viewed as pushing the same ball back to the top by overcoming all the impeding obstacles (epigenetic barriers), in the form of braes and curves, along the way. As mentioned in another section, epigenetics refers to any stable change in gene expression that is not due to changes in the DNA sequence but rather in the way this sequence is read and processed. In other words, even though every cell of our body has the same DNA information, the tissue-specific identity is determined by the way this DNA is accessible to transcription factors and how these transcription factors interact with other transcription factors and additional proteins including co-activators and co-repressors. The molecular basis of epigenetics is very complex but can be largely attributed to two main types of modifications: DNA methylation and chromatin remodeling [31].

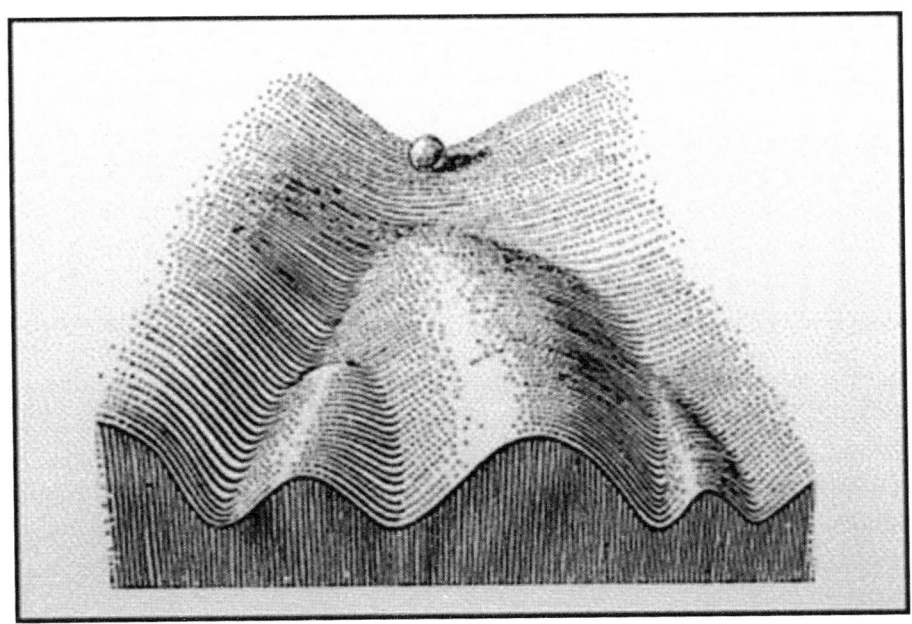

Waddington, C. H. 1957

Figure 9: The concept of epigenetic landscape by Waddington remains valid today despite it was proposed before the physical structure of genes and their role in heredity was unveiled. A cell represented by the ball can take a special path to become a particular fully differentiated cell. Modern molecular biology indicates that epigenetic does not involve the change of genetic sequence but shapes the path of the ball. Epigenetics may be defined as the study of any potentially stable and, ideally, heritable changes in gene expression or cellular phenotype that occurs without changes in Watson-Crick base-pairing of DNA.

DNA methylation has been more extensively studied and consists of the addition of a methyl group to the number 5 carbon of the pyrimidine ring, typically in a CpG dinucleotide context. This is mediated by a family of DNA methyltransferase enzymes (DNMTs). In somatic cells, approximately 40–60% of all CpGs are methylated, and this is a key element in the formation and maintenance of tissue-specific heterochromatin (a condensed form of the chromatin, which is therefore silent) [31]. One extended criterion for

167

efficient nuclear reprogramming is the demethylation of the proximal promoters of ES cell–specific transcription factors (i.e.,,Oct4 and Nanog), which otherwise are heavily methylated in somatic cells [19, 20]. Notably, genome-wide comparative study of hES and iPS cell-methylation profiles have also been reported and highlight significant differences between the two cell types, some of which are atributable to epigenetic memory of the donor cell type [32]. Recent studies show that the Tet family members are the major enzymes to mediate DNA demethylation [33]. Besides their role in development and nuclear reprogramming, changes in DNA methylation play as well an important role in cancer and other diseases [32].

Chromatin is the sum of DNA and the proteins with which it associates, in particular the histones [31]. DNA is wrapped around histones in order to pack the large eukaryotic genomes into the reduced nuclear environment while allowing appropriate access to the trancriptional machinery. This organizes the DNA in repeating units of chromatin: the nucleosomes. The nucleosome core particle consists of approximately 146 base pairs of DNA wrapped around a <u>histone</u> octamer consisting of two copies of the core histones <u>H2A</u>, <u>H2B</u>, <u>H3</u>, and <u>H4</u>. The covalent modification of the core histones modulates genome function either allowing or repressing transcription (Figure 10. Covalent chromatin modifications). One very important modification is histone methylation, which occurs on arginine and lysine residues. Other modifications include phosphorylation, acetylation, sumoylation, and ubiquitination. Currently both arginine and lysine methyl transferases and demethylases have been identified that have preference for one or another type of histone modification (for example, mono-, di-, and trimethylation). As a result, the complexity is immense. Histone acetyltransferases and histone deacetylases are also essential chromatin regulators but their role in ES cells has so far received less attention. Differentiation of ES cells triggers global changes in histone modifications and a transition to a transcriptionally less-permissive chromatin state [34, 35] (Figure 11. Chromatin structure of ES cells vs fully differentiated cells). For example, there is a global decrease in H3K4me3 (histone H3 trimethylated in lysine 4, a mark of transcriptional activation) and an elevation of H3K9 methylation (H3 lysine 9, a mark of transcriptional repression). Interestingly, many tissue-specific genes in undifferentiated ES cells also bear the so-called bivalent mark in undifferentiated ES cells, which consist of the repressive modification H3K27me3 (H3 trimethylated in lysine 27) and the activating H3K4me3. Genes marked by these bivalent domains are kept in a semi-permissive transcriptional state, which means that they are silent but posed for activation upon induction [35].

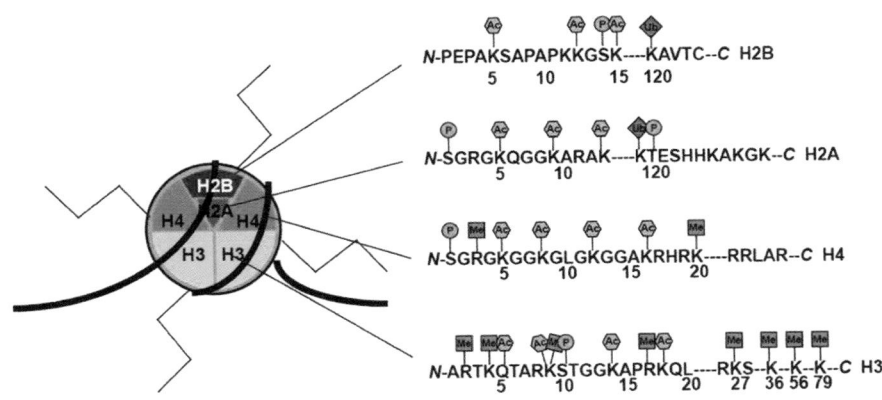

Figure 10: Different types of histone covalent modifications (Ac=acetylation, Me=methylation, P=phosphorylation, Ub=ubiquitination) have a profound impact on chromatin structure and gene expression.

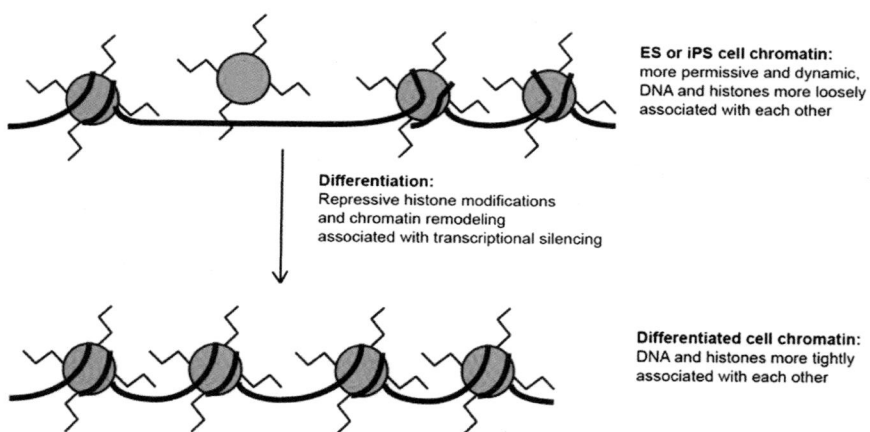

ES or iPS cell chromatin:
more permissive and dynamic,
DNA and histones more loosely
associated with each other

Differentiation:
Repressive histone modifications
and chromatin remodeling
associated with transcriptional silencing

Differentiated cell chromatin:
DNA and histones more tightly
associated with each other

Figure 11: According to the model, the chromatin of ES cells which is in an open hyperdynamic status progressively becomes more restricted upon tissue specific differentiation.

The other big family of chromatin regulators is the ATP-dependent chromatin remodeling complexes [36]. ATP-dependent chromatin remodeling complexes are specialized multi-protein machines that utilize ATP to non-covalently restructure, mobilize, or eject nucleosomes to regulate access to the DNA (Figure 12. ATP-dependent chromatin remodeling). There are four major families based on the catalytic subunit and the composition of the complexes: SWI/SNF, ISWI, Ino80, and Mi-2. Studies of these complexes have shown that the nucleosomal ATPase ISWI takes an essential role in nuclear reprogramming [37], and Chd1 of the Mi-2/CHD family is involved in both ES cell pluripotency and iPSC generation [38] and SWI/SNF components are essential for both ES cell pluripotency, tissue-specification, and iPSC generation [39-41]. ES cell-enriched subunits BAF250a and Brg1 are important to maintain ES cell pluripotency and self-renewal, whereas BAF60c, a cardiac tissue–specific SWI/SNF subunit, is sufficient to drive mesodermal cells into cardiac cells by working with a cardiac transcription factor GATA4 [42]. These studies together with a large body of work published earlier suggest that the dynamic composition and relative abundance of each individual component of the SWI/SNF complexes can form a steering force to determine either self-renewal or differentiation [43]. Indeed, a recent study in human ES cells identified the presence of SWI/SNF and low nucleosome density at both active and poised (bivalent) enhancers that are consistent with the essential function of the complex [44]. The detailed molecular mechanisms, such as how these chromatin remodeling complexes open up or close chromatin structure to drive cell fate change, await further investigation.

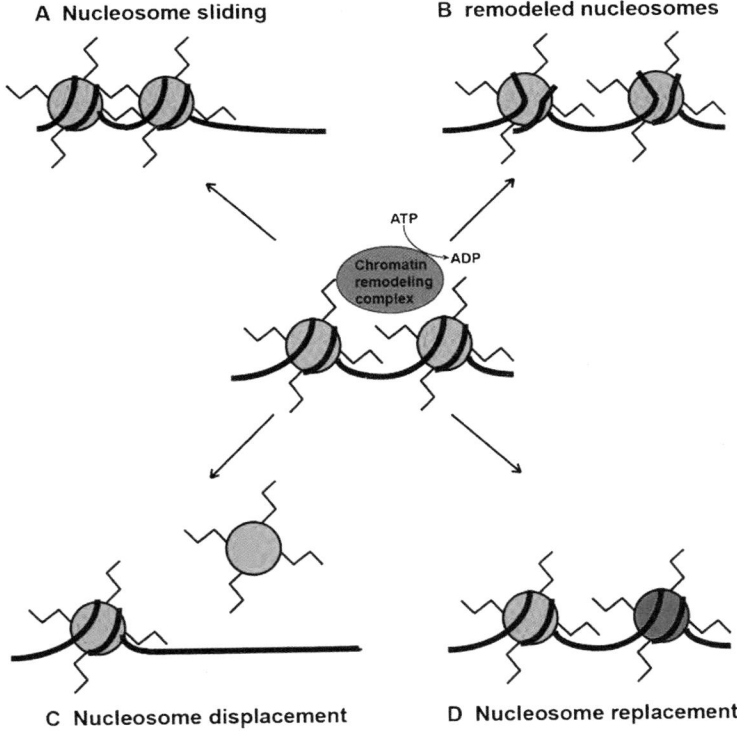

Figure 12: Chromatin remodeling complexes are a group of DNA-translocating motors that utilize the energy of ATP hydrolysis to change the contacts between histones and DNA. Four different models for their enzymatic function are proposed based mostly on in vitro studies: remodelers can (A)

mediate nucleosome sliding, in which the position of a nucleosome on the DNA changes; (B), create a remodeled state, or modified state, in which the DNA becomes more accessible but histones remain bound; (C) completely dissociate histones from DNA; and (D) replace histone with a variant histone.

6. Perspectives

SCNT and the direct reprogramming by defined factors to generate ES or iPS cells have revolutionized the stem cell field [45]. Even though Oct4, Sox2, Klf4, and c-Myc are able to achieve complete reprogramming, it has become obvious that other proteins or small molecules can replace some of these factors. To quote some examples: Thomson and colleagues used Oct4, Sox2, Lin28, and Nanog to produce human iPS cells [46]; the transcription factor Esrrb has been shown to replace Klf4 [47]; and use of a Tgfb inhibitor (Tgfb is present at high concentration in fetal bovine serum used for ES cell tissue culture) can substitute Sox2 or c-Myc [48, 49]. Once the epigenetic events that transform a somatic cell into an iPS cell are better understood, a combination of the most suitable cell type, the right tissue culture conditions, and chemical inhibitors may be sufficient to generate high quality iPS cells from any cell types without safety concerns. In this regard, it has been reported that microRNA mimics can produce mouse and human iPSCs [50].

One major obstacle in regenerative medicine is to generate pure fully committed tissue-specific progenitor cells or mature cells for cell-based therapy. Therefore, instead of converting iPS cells into the required cell type, lineage reprogramming that converts one mature cell type into another may be desirable in at least some instances (Figure 13. ES cell differentiation, direct reprogramming (iPS cells), and trans-differentiation in the epigenetic landscape). In fact, three transcription factors have been discovered to convert non-insulin producing pancreas cells to insulin-producing endocrine cells in mouse pancreas [51], and two factors appear sufficient to convert myoblasts into adipocytes [52]. More recent studies have also produced different lineages including neurons [53, 54] and heart-like cells by over-expressing tissue specific transcription factors [55]. Therefore, it is tempting to speculate that combined with appropriate epigenetic modulators, any tissue-specific cell lines can be generated in the Petri dish without the need to go first to the pluripotent stage (iPS cell).

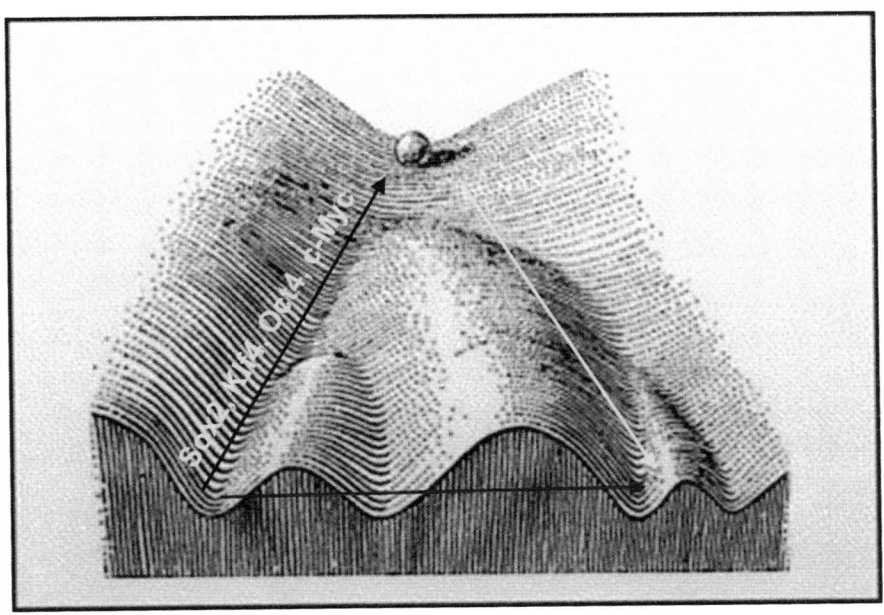

Figure 13: Yellow arrow indicates the differentiation of ES cells or iPSC into a given lineage (e.g. cardiomyocytes), blue arrow indicates the generation of iPSC from somatic cells (i.e. skin fibroblasts) using defined combinations of exogenous factors (e.g. Sox2, Klf4, Oct4, c-Myc), red arrow indicates the trans-differentiation of a somatic cell (fibroblast) into another somatic cell (cardiomyocyte) without the need to go to the pluripotent status first.

As a final consideration, it is likely that novel concepts arising from the study of ES or iPS cells will break through different disciplines and change the way unrelated diseases are perceived today. For example, the concept of cancer as a stem cell disease is becoming increasingly popular, and the emerging approaches to induce either pluripotency or differentiation may find parallelisms in this particular context [56, 57] (Figure 14. The stem cell theory of cancer).

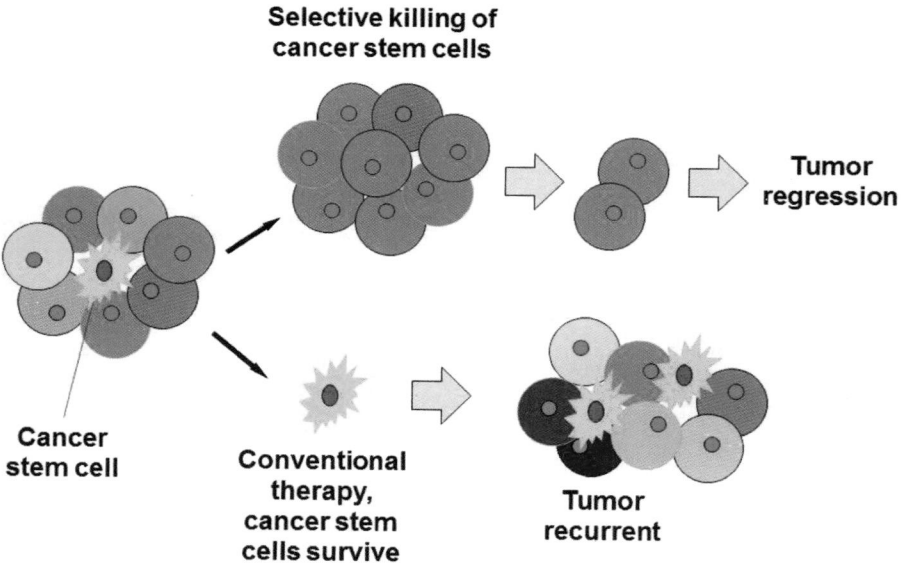

Figure 14: Cancers are normally composed of tumor cells exhibiting multiple stages of differentiation. The stem cell theory of cancer proposes the existence of cells with multipotent characteristics (cancer stem cells) inside cancers that are responsible for the sustained tumor growth after standard therapies as these cells are resistant to conventional treatments. On the other hand, approaches that only kill the cancer stem cells may prove to cure the cancer given that most tumor cells are partially differentiated and have lost the ability to self renew. Currently it is also debated whether these cancer stem cells are consequence of cancers acquiring stem cell characteristics during tumor progression or whether cancers can also originate from tissue specific progenitor cells.

References

1. Jones JM, Thomson JA. Human embryonic stem cell technology. *Semin Reprod Med* 2000; 18:219-223.

2. Nagy A, Gertsenstein, M., Vintersten, K., Behringer, R. Summary of mouse development. In: Nagy A, Gertsenstein, M., Vintersten, K., Behringer, R. ed. Manipulating the Mouse Embryo. New York: Cold Spring Harbor Laboratory Press 2003:31-139.

3. Odorico JS, Kaufman DS, Thomson JA. Multilineage differentiation from human embryonic stem cell lines. *Stem Cells* 2001; 19:193-204.

4. Evans MJ, Kaufman MH. Establishment in culture of pluripotential cells from mouse embryos. *Nature* 1981; 292:154-156.

5. Thomson JA, Itskovitz-Eldor J, Shapiro SS *et al.* Embryonic stem cell lines derived from human blastocysts. *Science* 1998; 282:1145-1147.

6. Thomson JA, Kalishman J, Golos TG *et al.* Isolation of a primate embryonic stem cell line. *Proc Natl Acad Sci U S A* 1995; 92:7844-7848.

7. Fernandez M, Pirondi S, Chen BL *et al.* Isolation of rat embryonic stem-like cells: a tool for stem cell research and drug discovery. *Dev Dyn* 2011; 240:2482-2494.

8. Brons IG, Smithers LE, Trotter MW *et al.* Derivation of pluripotent epiblast stem cells from mammalian embryos. *Nature* 2007; 448:191-195.

9.   Taylor CJ, Bolton EM, Bradley JA. Immunological considerations for embryonic and induced pluripotent stem cell banking. *Philos Trans R Soc Lond B Biol Sci* 2011; 366:2312-2322.

10.   Tang C, Lee AS, Volkmer JP *et al*. An antibody against SSEA-5 glycan on human pluripotent stem cells enables removal of teratoma-forming cells. *Nat Biotechnol* 2011; 29:829-834.

11.   Schwartz SD, Hubschman JP, Heilwell G *et al*. Embryonic stem cell trials for macular degeneration: a preliminary report. *Lancet* 2012.

12.   Biancotti JC, Benvenisty N. Aneuploid human embryonic stem cells: origins and potential for modeling chromosomal disorders. *Regen Med* 2011; 6:493-503.

13.   Wilmut I, Schnieke AE, McWhir J, Kind AJ, Campbell KH. Viable offspring derived from fetal and adult mammalian cells. *Nature* 1997; 385:810-813.

14.   Gurdon JB, Byrne JA. The first half-century of nuclear transplantation. *Proc Natl Acad Sci U S A* 2003; 100:8048-8052.

15.   Hochedlinger K, Jaenisch R. Nuclear reprogramming and pluripotency. *Nature* 2006; 441:1061-1067.

16.   Noggle S, Fung HL, Gore A *et al*. Human oocytes reprogram somatic cells to a pluripotent state. *Nature* 2011; 478:70-75.

17.   Takahashi K, Yamanaka S. Induction of pluripotent stem cells from mouse embryonic and adult fibroblast cultures by defined factors. *Cell* 2006; 126:663-676.

18.   Cole MF, Young RA. Mapping key features of transcriptional regulatory circuitry in embryonic stem cells. *Cold Spring Harb Symp Quant Biol* 2008; 73:183-193.

19. Okita K, Ichisaka T, Yamanaka S. Generation of germline-competent induced pluripotent stem cells. *Nature* 2007; 448:313-317.

20. Wernig M, Meissner A, Foreman R *et al.* In vitro reprogramming of fibroblasts into a pluripotent ES-cell-like state. *Nature* 2007; 448:318-324.

21. Zhao XY, Lv Z, Li W, Zeng F, Zhou Q. Production of mice using iPS cells and tetraploid complementation. *Nat Protoc* 2010; 5:963-971.

22. Rajarajan K, Engels MC, Wu SM. Reprogramming of Mouse, Rat, Pig, and Human Fibroblasts into iPS Cells. *Curr Protoc Mol Biol* 2012; Chapter 23:Unit23 15.

23. Zhou T, Benda C, Duzinger S *et al.* Generation of induced pluripotent stem cells from urine. *J Am Soc Nephrol* 2011; 22:1221-1228.

24. Li W, Wang X, Fan W *et al.* Modeling abnormal early development with induced pluripotent stem cells from aneuploid syndromes. *Hum Mol Genet* 2012; 21:32-45.

25. Park IH, Arora N, Huo H *et al.* Disease-specific induced pluripotent stem cells. *Cell* 2008; 134:877-886.

26. Esteban MA, Wang T, Qin B *et al.* Vitamin C enhances the generation of mouse and human induced pluripotent stem cells. *Cell Stem Cell* 2010; 6:71-79.

27. Banito A, Gil J. Induced pluripotent stem cells and senescence: learning the biology to improve the technology. *EMBO Rep* 2010; 11:353-359.

28. Pera MF. Stem cells: The dark side of induced pluripotency. *Nature* 2011; 471:46-47.

29. Esteban MA, Peng M, Deli Z *et al.* Porcine induced pluripotent stem cells may bridge the gap between mouse and human iPS. *IUBMB Life* 2010; 62:277-282.

30. Waddington CH. The Strategy of the Genes: a discussion of some aspects of theoretical biology. London: Geo Allen & Unwin 1957.

31. allis CDJ, T; Reinberg, D. Overview and concepts. In: Allis CD, Jenuwein, T., Reinberg, D. ed. Epgenetics. New York: Cold Spring Harbor Laboratory Press 2007:23-62.

32. Kim K, Doi A, Wen B *et al.* Epigenetic memory in induced pluripotent stem cells. *Nature* 2010; 467:285-290.

33. Veron N, Peters AH. Epigenetics: Tet proteins in the limelight. *Nature* 2011; 473:293-294.

34. Zwaka TP. Breathing chromatin in pluripotent stem cells. *Dev Cell* 2006; 10:1-2.

35. Chi AS, Bernstein BE. Developmental biology. Pluripotent chromatin state. *Science* 2009; 323:220-221.

36. Mohrmann L, Verrijzer CP. Composition and functional specificity of SWI2/SNF2 class chromatin remodeling complexes. *Biochim Biophys Acta* 2005; 1681:59-73.

37. Kikyo N, Wade PA, Guschin D, Ge H, Wolffe AP. Active remodeling of somatic nuclei in egg cytoplasm by the nucleosomal ATPase ISWI. *Science* 2000; 289:2360-2362.

38. Gaspar-Maia A, Alajem A, Polesso F et al. Chd1 regulates open chromatin and pluripotency of embryonic stem cells. *Nature* 2009; 460:863-868.

39. Gao X, Tate P, Hu P, Tjian R, Skarnes WC, Wang Z. ES cell pluripotency and germ layer formation require the SWI/SNF chromatin remodeling component BAF250a. *Proc Natl Acad Sci U S A* 2008; 105:6656-6661.

40. Ho L, Ronan JL, Wu J et al. An embryonic stem cell chromatin remodeling complex, esBAF, is essential for embryonic stem cell self-renewal and pluripotency. *Proc Natl Acad Sci U S A* 2009; 106:5181-5186.

41. Singhal N, Graumann J, Wu G et al. Chromatin-Remodeling Components of the BAF Complex Facilitate Reprogramming. *Cell* 2010; 141:943-955.

42. Takeuchi JK, Bruneau BG. Directed transdifferentiation of mouse mesoderm to heart tissue by defined factors. *Nature* 2009; 459:708-711.

43. Wu JI, Lessard J, Crabtree GR. Understanding the words of chromatin regulation. *Cell* 2009; 136:200-206.

44. Rada-Iglesias A, Bajpai R, Swigut T, Brugmann SA, Flynn RA, Wysocka J. A unique chromatin signature uncovers early developmental enhancers in humans. *Nature* 2011; 470:279-283.

45. Wilmut I, Sullivan G, Taylor J. A decade of progress since the birth of Dolly. *Reprod Fertil Dev* 2009; 21:95-100.

46. Yu J, Vodyanik MA, Smuga-Otto K et al. Induced pluripotent stem cell lines derived from human somatic cells. *Science* 2007; 318:1917-1920.

47. Feng B, Jiang J, Kraus P *et al*. Reprogramming of fibroblasts into induced pluripotent stem cells with orphan nuclear receptor Esrrb. *Nat Cell Biol* 2009.

48. Ichida JK, Blanchard J, Lam K *et al*. A Small-Molecule Inhibitor of Tgf-beta Signaling Replaces Sox2 in Reprogramming by Inducing Nanog. *Cell Stem Cell* 2009.

49. Li R, Liang J, Ni S *et al*. A mesenchymal-to-epithelial transition initiates and is required for the nuclear reprogramming of mouse fibroblasts. *Cell Stem Cell* 2010; 7:51-63.

50. Miyoshi N, Ishii H, Nagano H *et al*. Reprogramming of mouse and human cells to pluripotency using mature microRNAs. *Cell Stem Cell* 2011; 8:633-638.

51. Zhou Q, Brown J, Kanarek A, Rajagopal J, Melton DA. In vivo reprogramming of adult pancreatic exocrine cells to beta-cells. *Nature* 2008; 455:627-632.

52. Hu E, Tontonoz P, Spiegelman BM. Transdifferentiation of myoblasts by the adipogenic transcription factors PPAR gamma and C/EBP alpha. *Proc Natl Acad Sci U S A* 1995; 92:9856-9860.

53. Pang ZP, Yang N, Vierbuchen T *et al*. Induction of human neuronal cells by defined transcription factors. *Nature* 2011; 476:220-223.

54. Yoo AS, Sun AX, Li L *et al*. MicroRNA-mediated conversion of human fibroblasts to neurons. *Nature* 2011; 476:228-231.

55. Ieda M, Fu JD, Delgado-Olguin P *et al*. Direct reprogramming of fibroblasts into functional cardiomyocytes by defined factors. *Cell* 2010; 142:375-386.

56. Takebe N, Harris PJ, Warren RQ, Ivy SP. Targeting cancer stem cells by inhibiting Wnt, Notch, and Hedgehog pathways. *Nat Rev Clin Oncol* 2011; 8:97-106.

57. Clevers H. The cancer stem cell: premises, promises and challenges. *Nat Med* 2011; 17:313-319.

Chapter 7

Genetic Manipulation of Adult Stem Cells: Methods and Applications

Carl Gregory
Institute for Regenerative Medicine at Scott and White Hospital, Texas A and M Health Science Center, Module C, 5701 Airport Road, Temple, TX, USA.

1. Introduction

For decades, stem and progenitor cells from a wide variety of adult tissues (for example, bone marrow and neuronal tissue) have been identified and cultured with a view to harnessing their substantial promise for the treatment of countless human disorders. Adult stem (AS) cells have received special attention for the treatment of chronic genetic diseases such as osteogenesis imperfecta, severe combined immune deficiency, haemophilia, and cystic fibrosis, for which there is no cure. In the case of genetic diseases, there is an obvious drawback in that autologously derived stem cells from the affected individual will continue to harbor the deficit that causes the disease. Of course, in some cases, immunologically-compatible donors may be found, but autologous transplantation of genetically repaired cells would be immensely beneficial. Furthermore, the incorporation of foreign DNA into the genome of the donor cells provides the opportunity for surveillance of the engrafted cells by PCR or additional means if proteins are introduced that facilitate immunological tagging. Experimentally, the ability to genetically alter adult stem cells has revolutionized the field of regenerative medicine in that engrafted cells can be visualized histologically, or more recently, in living, intact animals by means of fluorescent or bioluminescent imaging. Given that unmanipulated stem cells from mammalian sources do not possess proteins with suitable fluorescent or bioluminescent properties, genetic manipulation is essential in these types of studies. With these considerations in mind, there is a clear necessity for the advancement of genetic manipulation techniques for AS cells. In retrospect, genetic alteration of primary adult stem cells should be a straightforward affair since the technology has been widely utilized for some decades ; however, adult stem cells have diverse unique properties in culture that in many cases require special provision. For example, mesenchymal stem (MS) cells are extremely sensitive to seeding density and population doublings. Therefore, transfection or transduction methods that require high-density cultures or result in low efficiencies are usually unsuitable for this application. This chapter reviews a selection of commonly employed methodologies for the genetic alteration of mammalian cells with particular emphasis on their suitability for the modification of AS cells. Given that adult stem or progenitor cells can be cultured from nearly every tissue in the mammalian organism, examples will be provided for the most common classes of

AS cells, namely MS cells, hematopoietic stem (HS) cells, and neural progenitor (NS) cells.

2. Transfection of Nucleic Acids

Transfection is the term given to any method that results in the direct transfer of DNA to a eukaryotic cell in the absence of internalization that is facilitated by viral means. There are numerous means by which the DNA can be functionally incorporated into the cell resulting in various levels of efficiency and longevity of gene expression, but the most common modern means of transfection are electroporation and lipofection. Electroporation involves the application of a strong electric pulse to the cells, resulting in a vigorous redistribution of the ionic solutes around the lipid bilayer of the plasma membrane. This results in localized rearrangement of the lipids in the bilayer, causing the transient formation of pores, which allow DNA or other materials to enter the cytoplasm [75, 73, 74]. Lipofection belongs to a family of transfection strategies that involve coating of the negatively charged nucleic acids in a cationic carrier prior to exposure and uptake by the cells [18]. The earliest protocols involved mixing the target DNA with calcium phosphate rather than lipid complexes, [18, 56, 33] but more recent strategies have adopted complex cationic lipids such as DOTMA (N-[2,3, dioleyloxy]-propyl]-N,N,N-triethylammonium chloride, Lipofectin) or DOPE (dioleoylphosphotydylethanolamine) which have a far higher efficiency than calcium phosphate [21, 20]. The mechanism of carrier–nucleic acid complex uptake is not completely understood, but it has been suggested that passage through the plasma membrane may occur by direct fusion of the lipid components of the complex with the bilayer or by endocytosis. Once inside the cell, and if they can escape lysosomal degradation, the carrier–nucleic acid complexes are free to diffuse across the nuclear envelope and participate in gene expression.

The stability of the transfection determines the longevity of gene expression. Generally, if the DNA becomes incorporated into the genome of the cell, gene expression will persist, and the transfection may be regarded as stable. It should be noted, however, that constitutive expression of a protein that substantially perturbs cellular physiology can lead to selection of clones with a very low level of expression, or even complete arrest of target gene expression. Transient transfection occurs when the plasmid DNA becomes active in the nucleus, but does not integrate into the genome of the host cell. As the name implies, the activity of the construct is limited in most cases to a few days, depending on the proliferative rate of the culture and the half-life of the nucleic acid construct. Generally, the highest proportion of DNA uptake events leads to transient transfections with a far lower frequency, resulting in genomic integration and stable gene transfer. The proportion of transient-to-stable DNA uptake events can be increased in some cases by employing electroporation, rather than lipofection, or by linearizing the DNA. Although the utility of transient transfectants is substantially reduced, the resultant cells are easily generated and are useful for short-duration studies such as reporter gene assays.

The use of stably transfected AS cells for in vivo studies has been limited to a handful of instances because the current viral technology is better suited to the application. However, there are some examples where stably transfected AS cells have been genetically altered to express tracking proteins or experimental therapeutic proteins in MS cells [48, 43, 78], HS cells [47], and NPCs [41]. Direct nucleic acid transfection has most utility in short duration reporter gene and short interfering RNA (siRNA) blocking assays where transient transfection is acceptable. In reporter gene assays, stem cells are transfected with a DNA construct consisting of a promoter of interest coupled to a reporter gene encoding an enzyme that can be quantitatively assayed. In most instances, the reporter gene is fluorescent or has the capacity to generate luminescence in the presence of suitable substrate. When the DNA construct is transfected, and the cells are perturbed, the effect on the test promoter can be directly quantified by measuring the level of the reporter enzyme produced. The utility of this approach is very well exemplified by elegant studies performed on the T cell factor/lymphoid enhancer factor (TCF/LEF) promoter module activated by WnT signaling during growth and development [36]. Transient transfection techniques have also permitted fast and convenient transfer of siRNA into AS cells for gene expression knock down experiments. In such studies, cultured cells are transfected with short sequences of RNA that are complimentary to the transcript of interest (the siRNA). In a process termed *RNA interference* (RNAi) the siRNA forms a duplex with the transcript, blocking translation and targeting its degradation [44, 34]. The net result is specific inhibition of the expression of the target gene. Such gene expression knock down experiments have contributed substantially to our knowledge of stem cell biology [58, 80]. Furthermore, because transient lipofection is particularly suited to high-throughput screening, large-scale functional genomics is possible using siRNA molecules directed to thousands of different transcripts [69].

3. Adenoviral Transduction

Adenoviruses were discovered in the early 1950s as part of the initiative to find a cure for the common cold. There are at least fifty known serotypes, which generally target gastrointestinal, respiratory, and ocular epithelial tissue. The virions are approximately 80 nm diameter icosahedrons comprised of three proteins known as the fiber, penton base, and hexon. Fiber and penton base proteins are necessary for receptor binding and cell internalization and protrude from the capsid wall, which is composed of the hexon protein. The wild-type viral genome consists of a single double-stranded DNA molecule approximately 37 kilobases long, encoding about seventy genes. Viral infection is a complex process initiated by docking of the virion to the coxackievirus and adenovirus receptor (CAR) on the plasma membrane of the host cell via the fiber and penton base. Internalization occurs and after escape from the endocytotic endosome, the virus disassembles as it migrates to the nucleus. In the nucleus, gene transcription occurs from the viral genome without integration into the host chromosomes. Rather, it exists in the nucleus as a circular piece of independent DNA. In the 1970s and 1980s, the utility of adenoviruses as gene therapy tools became apparent and much effort was devoted to generating recombinant versions. In particular, similarities between bacterial plasmids and the circular conformation of the genome in infected cells were exploited, and bacterial vectors with adenoviral properties were born [24, 23]. Because wild-type adenoviruses ultimately destroy the cell upon infection, replication deficient viral vectors were generated, which lacked a critical component of the replication machinery (the *E1A* gene) as well as superfluous components to increase the size limit of the recombinant DNA that could be included in the construct. To date, about 8 kilobases of DNA can be incorporated into adenoviral vectors which have the capacity to infect a given cell once, without subsequent initiation of the replicative and lytic cycle. Because the adenoviruses generated by recombinant vectors are replication deficient, complementation of the *E1A* gene

187

is necessary to facilitate viral replication. This is achieved by utilizing a standard host cell line stably transfected with the *E1A* gene, known as the 293 packaging cell line. The adenoviral vector can be transfected into the packaging line and undergo viral replication via the activity of the host copy of the *E1A* gene. Resultant virions from the packaging culture contain only the viral vector, lack a copy of the *E1A* gene, and can therefore enter a host cell once, without subsequent replication.

Adenoviruses have been used widely in adult stem cell research as an experimental tool and also a potential therapeutic agent [32]. Although gene expression is transient, due to the absence of genomic integration, the probability of insertional mutagenesis is virtually zero. Furthermore, transient expression of some therapeutic genes is desirable. For instance, recombinant expression of the *MDR1* gene causes improved HSC expansion in culture, but can cause leukemia when engrafted in vivo [11]. At first glance, adenoviruses therefore appear to be safe candidates for pre-modification of transiently acting therapeutic stem cell preparations, but there are drawbacks. Because viral proteins continue to be expressed in the host cells, and humans are frequently exposed to wild-type adenoviral insults, cell preparations exposed to adenoviral vectors can illicit similar immune responses that occur when the virus is directly administered for gene therapy [26, 55]. Also, some stem cell preparations are poorly transduced by adenoviruses due to low densities of the CAR receptor necessitating extremely high concentrations of virus. This is particularly true for human MS cells and HS cells [32], but in some cases this problem can be partially solved by modifying the coat proteins of the vector or providing a chemical additive to increase the efficiency of binding to the target cells [25, 67, 32, 7]. At any rate, the utility of adenoviruses in basic AS cell research is substantial, with effective gene transfer achieved in MS cells and stromal cells [14, 68], HS cells [30, 42], endothelial and cardiac progenitors [71], and NS cells [5].

4. Adeno-Associated Viruses

Adeno-associated viruses (AAV) are parvoviruses, which exist as satellite viruses of other human viruses such as adenoviruses, and thus require co-infection with the associated viruses. AAVs are smaller than adenoviruses (approximately 23 nm in diameter) but are similar in shape. The AAV genome is also slightly smaller at 4–5 kilobases and single stranded. At each end of the viral genome is a palindromic sequence known as the inverted terminal repeat (ITR). The ITR imparts a unique characteristic in that it facilitates selective integration into a locus on chromosome 19 of the human genome (designated AAVS1). Random insertions do occur, but these events are rare. The utility of this characteristic is clear in that it provides for directed integration of a construct, minimizing concerns of insertional mutagenesis. Adeno-associated vectors are prepared in a similar way to adenoviral vectors, using a cell- and plasmid-based packaging system [10]. For preparation, a packaging cell line is first infected by adenovirus or herpes simplex virus (HSV). Shortly after infection, the cells are transfected with two plasmids: one containing the gene of interest and the ITRs necessary for packaging as well as integration; and a second plasmid that lacks the ITRs but contains the remaining necessary viral genes for replication. The genes on both constructs complement each other, producing live virus, but only the plasmid containing the gene of interest and the ITRs is packaged in the virion, resulting in a replication deficient virus. After replication, the AAV can be purified from the resultant cell lysate and contaminating adenoviral or HSV particles by density

189

gradient centrifugation. Because AAVs readily transduce many types of human cells and are very restricted in genomic integration, they are excellent candidates for gene modification in AS cells and, to date, are the only DNA-based viral vector not associated with oncogenesis. However, AAVs share some of the immunological drawbacks of adenoviruses and can accommodate only about 5 kilobases of DNA, limiting their use to simple genes. In addition, the efficiency of AAV transduction is limited by expression of the receptors *heparan sulphate proteoglycan (HSP)*, *integrin alphaVbeta5*, and *fibroblast growth factor receptor 2 (FGFR)* on the plasma membrane [64, 65, 51].

The potential use of AAVs for genetic manipulation of AS cells has been explored with controversial results. In the case of HS cells, the presence of the co-receptors seems to be a critical factor in the success of transduction, but, generally, efficiencies are very low [62]. In NS cells and MS cells, however, the results have been somewhat more promising with reports of stable transgene expression occurring at relatively high frequency with AAV variants [72, 40, 46, 76], most notably in Parkinson's disease and orthopedic studies.

5. Lentiviruses

As the name suggests (*lenti-* means "slow" in Latin), lentiviruses belong to a family of complex integrating retroviruses which induce slowly manifesting chronic diseases that are usually associated with macrophage infection. There are five subgroups or serogroups that are subdivided based on the host they predominantly infect: primate (e.g., the human immunodeficiency virus, HIV), bovine (e.g., the bovine immunodeficiency virus), equine (e.g., the equine infectious anemia virus), feline (e.g., the feline immunodeficiency virus), and ovine (e.g., the Visna virus). The virions are spherical, at an average of 90 nm in diameter with protein protrusions emanating from the outer lipid and protein-based matrix. The protein protrusions are glycoproteins (GP120, GP41), and GP120 is responsible for docking with target cells. Within the exterior matrix is a core capsid, where two molecules of RNA encode a diploid viral genome. Accompanying the RNA within the capsid is a variety of polymerases, proteases, and RNAses necessary for viral integration and propagation. The viral genome is about 10 kilobases long and produces three main transcripts: *gag*, *pol*, and *env*. The translation products are fusion proteins which are subsequently cleaved into eight separate proteins by the viral protease. Assembled virions consist of three structural proteins to build the inner and outer capsids, two docking glycoproteins, a protease to process the transcripts, a reverse transcriptase to convert the RNA genome to DNA for genomic integration, and an integrase to mediate genomic integration of the reverse transcribed viral genome. Lentiviruses infect dividing and nondividing cells by docking with receptors on the host cell membrane through specific interaction by the GP120 protein. Upon docking, GP41 triggers a membrane fusion event between the plasma membrane of the target cell and the viral matrix that facilitates access to the cell cytoplasm. The viral RNA is reverse transcribed to DNA and translocated to the nucleus where it is integrated into the host cell genome by the action of the viral integrase. Viral propagation occurs through transcription and translation of the integrated viral genome. Viral RNA is specifically packaged into functional virions through the action of a packaging signal in the viral genome known as the ψ sequence.

Lentiviruses have received great attention over the last few decades not only because they are the family to which HIV belongs, but also due to their great potential as a gene transfer tool. The virions have the ability to transduce nondividing quiescent cells, which has been the major limitation of older retroviruses [22]. Given that many stem cell lines do not substantially divide and in some cases react poorly to direct DNA transfection, lentiviral vectors provide an excellent method for stable gene transfer. Furthermore, when compared to adenoviruses, AAVs, and older retroviral tools such as the Moloney murine leukemia virus (MoMLV), lentiviruses have extremely low toxicity and do not elicit a substantial inflammatory response [31]. The generation of a safe and effective strategy to harness the benefits of the lentivirus has been substantially aided by the identification of the Ψ sequence, which enables specific packaging of RNA strands into virions, and the nucleic acid sequences known as long terminal repeats (LTRs), which permit integration of the reverse-transcribed viral genome into the genomic DNA of the host. In the absence of additional components, a virion harboring vector containing only these components could integrate once, but would lack the necessary genes to express further functional virions. Therefore, this virion could be regarded as replication deficient. In practice, replication-deficient lentiviral virions are generated in a similar manner to adenoviruses. The most modern and theoretically safest systems employ three or four DNA plasmids: one contains the gene of interest, LTRs, and Ψ sequence (sometimes known as the transfer vector); the remaining plasmids encode structural and functional components, but lack LTRs or Ψ sequences. The plasmids are transiently transfected simultaneously (co-transfected) into a packaging cell line, and those cells that received the full complement of plasmids proceed to express virion components, assemble virions, and package only the transfer vector. Replication-deficient virions can then be purified from the growth media of the cultured packaging cell line [15].

Lentiviral vectors are extremely versatile and have been used extensively for tracking engrafted AS cells from a variety of tissues. Because the integration of a given sequence is random, single clones and their progeny can be identified and quantified in vitro. This is especially useful in the hematopoietic field where HS cells expand from single cell derived clones and differentiate into a variety of different kinds of blood cell. Studies have been conducted where HS cells were each randomly tagged by a single lentiviral integration and injected in vivo. When the blood cells are re-extracted, they can be analyzed by PCR, and vast numbers of progeny can be tracked back to single HS cells [61, 59]. Because HS cells have substantial longevity in vivo after implantation and circulate widely, there has also been much interest in introduction of potentially therapeutic genes by lentiviral transduction [8, 19, 38, 57]. Some noteworthy examples include potential treatment for beta-thalassemias [54], Gaucher disease [35], metachromatic leukodystrophy [6], and sickle cell anemia [49]. Lentiviral vectors are also extremely efficient at stably modifying MS cells [79], suggesting that lentivirally modified MS cells may also be useful tools for the delivery of therapeutic proteins. Worthy of note is the lentivirally enhanced expression of bone repair proteins by MS cells such as osteoprotegerin [52] or bone morphogenic protein 2 [63, 28], treatment of nonobese diabetic severe combined immunodeficient mucopolysaccharidosis type VII in experimental animals by expression of the lysosomal enzyme beta-glucuronidase [45], or expression of a protein that induces apoptosis in some tumor cells [37]. Lentiviruses have been successful in stably modifying NS cells, too [12], suggesting that incurable neuronal diseases such as Parkinsonism could be treated by the stem-cell mediated delivery of therapeutic proteins [17].

Although promising, lentiviral transduction is not without its safety concerns. Because the viral construct integrates into the host genome, there is a probability for insertional mutagenesis of the host genes. Many such instances are not detected because loss of an essential protein frequently leads to cell death, but insertional events that lead to onco-genesis, known as insertional oncogenesis, have been re-ported with retrovirally modified HS cells in human subjects [2, 4, 27]. Insertional oncogenesis may occur at a frequency higher than other retroviruses because lentiviral integration sites occur more frequently in areas of high genomic activ-ity [60]. It appears, therefore, that more investigation is nec-essary to facilitate control over lentiviral insertion sites.

6. Less Common Methods for the Genetic Modification of ASC

6.1 Retroviral constructs

Before the advent of lentiviral and adenoviral vectors, on-cogenic retroviruses were frequently employed to stably modify mammalian cells. One of the most common retro-viral vectors was based on the Moloney murine leukemia virus. As the name suggests, the vector is not without its dis-advantages, requiring dividing cells for infection and caus-ing insertional oncogenesis in host HS cells [1]. In a clinical trial for treatment of x-linked severe combined immune de-ficiency, the modified HS cells caused leukemia in human recipients [2, 4, 14, 27].

6.2 Transposons

Transposons are mobile elements of nonviral DNA that can become inserted into the genome of mammalian cells. Due to their ability to efficiently insert into the genome, they have attracted much interest as genetic modification tools.

Transposon-based vectors, have been successfully used to stably modify HS cells [29], but there have been reports that insertional mutagenesis may cause sarcomas from murine MS cells [66].

6.3 Recombinant zinc finger nucleases

As discussed, one of the major drawbacks of stable gene manipulation in mammalian cells is the probability of insertional mutagenesis. Zinc finger nucleases are a class of restriction endonucleases that cut DNA at designated sequences with excellent specificity [13]. Although the technology is still in its infancy, it is possible to modify the amino acid sequence of the DNA binding domain of the nuclease to tailor design the point of DNA cleavage and transiently express the nuclease within a mammalian cell. The nuclease would be predicted to cut genomic DNA at the designated site [16, 76]. Transfection of an additional strand of DNA encoding the gene of interest with ends homologous to flanking sequences of the nuclease site could result in site-specific homologus recombination by the host cells' own recombination machinery [50]. There are a few recent examples where this technology has been applied to AS cells, including induced chromosomal translocations in MS cells derived from embryonic stem cells [9] and specific gene targeting, and in a human HSC-like cell line [39]. It seems, therefore, that zinc finger nucleases may represent a potential solution for the problem of random genomic integration. Indeed, use of integration-deficient lentiviral vectors that infect cells without integration, in conjunction with zinc finger nucleases, has resulted in directed gene modification in stem cells [39]

7. Final Remarks

Adult stem cells have great potential for the treatment of a wide range of diseases and injuries that are presently incurable. The cells themselves have been shown to have remarkable inherent curative properties, but in some cases supplementation or recovery of their function may be necessary, especially in genetic deficiencies. Furthermore, there is great scientific value in genetically modifying a stem cell to express a detectable marker or reporter protein, especially if the cells are to be tracked and quantified in vivo. For this purpose, there are a number of commonly employed strategies for genetic modification of AS cells. Genomic integration results in the most stable outcome, but this is fraught with the problem of insertional mutagenesis, which can cause tumors. However, alternative strategies that do not result in genomic integration lead to transient gene expression. Therefore, site-directed genomic integration is the long term goal, which may be achieved in the future by the use of zinc finger nucleases in conjunction with conventional gene transfer methods.

References

1.   Barquinero, J., Eixarch, H., and Pérez-Melgosa, M. 2004. Retroviral vectors: new applications for an old tool. *Gene Ther.* 11:S1, S3–9.

2.   Baum, C., Dullmann, J., Li, Z., Fehse, B., Meyer, J., Williams, D. A., and von

3.   Kalle, C. 2003. Side effects of retroviral gene transfer into hematopoietic stem cells. *Blood* 101:2099–2114.

4. Baum, C., von Kalle, C., Staal, F. J., Li, Z., Fehse, B., Schmidt, M., Weerkamp, F., Karlsson, S., Wagemaker, G., and Williams, D. A. 2004. Chance or necessity? Insertional mutagenesis in gene therapy and its consequences. *Mol. Ther.* 9:5–13.

5. Bertram, C. M., Hawes, S., Egli, S., Peh, G. S., Dottori, M., Kees, U. R., and Dallas, P. 2009. Effective adenovirus medi-ated gene transfer into neural stem cells derived from hu-man embryonic stem cells. *Stem Cells Dev.* Epub Jul 13 2009

6. Biffi, A., De Palma, M., Quattrini, A., Del Carro, U., Amadio, S., Visigalli, I., Sessa, M., Fasano, S., Brambilla, R., Marchesini, S., Bordignon, C., and Naldini, L. 2004. Correction of metachro-matic leukodystrophy in the mouse model by transplanta-tion of genetically modified hematopoietic stem cells. *J. Clin. Invest.* 113:1118–1129

7. Bosch, P and Stice, S. L. 2007. Adenoviral transduction of mesenchymal stem cells. *Methods Mol. Biol.* 407:265–274.

8. Brenner, S. and Malech, H. L. 2003. Current developments in the design of onco-retrovirus and lentivirus vector systems for hematopoietic cell gene therapy. *Biochim. Biophys. Acta.* 1640:1–24.

9. Brunet, E., Simsek, D., Tomishima, M., DeKelver, R., Choi, V. M., Gregory, P., Urnov, F., Weinstock, D. M., and Jasin, M. 2009. Chromosomal translocations induced at specified loci in hu-man stem cells. *Proc. Natl. Acad. Sci. USA* 106:10620–10625.

10. Büning, H., Perabo, L., Coutelle, O., Quadt-Humme, S., and Hallek, M. 2008. Recent developments in adeno-associated virus vector technology. *J. Gene Med.* 10:717–733.

11. Bunting, K. D., Galipeau, J., Topham, D., Benaim, E., and Sor-rentino, B. P. 1998. Transduction of murine bone marrow cells

with an MDR1 vector enables ex vivo stem cell expansion, but these expanded grafts cause a myeloproliferative syndrome in transplanted mice. *Blood* 92:2269–2279.

12. Capowski, E. E., Schneider, B. L., Ebert, A. D., Seehus, C. R., Szulc, J., Zufferey, R., Aebischer, P., and Svendsen, C. N. 2007. Lentiviral vector-mediated genetic modification of human neural progenitor cells for ex vivo gene therapy. *J. Neurosci Methods* 163:338–349.

13. Carroll D. 2008. Progress and prospects: zinc-finger nucleases as gene therapy agents. *Gene Ther.* 15:1463–1488.

14. Cavazzana-Calvo, M., Hacein-Bey, S., de Saint Basile, G., Gross, F., Yvon, E., Nusbaum, P., Selz, F., Hue, C., Certain, S., Casanova, J. L., Bousso, P., Deist, F. L., and Fischer, A. 2000. Gene therapy of human severe combined immuno-deficiency SCID.-X1 disease. *Science* 288:669–672.

15. Cheng, S. L., Lou, J., Wright, N. M., Lai, C. F., Avioli, L. V., and Riew, K. D. 2001. In vitro and in vivo induction of bone formation using a recombinant adenoviral vector carrying the human BMP-2 gene. *Calcif. Tissue Int.* 68:87–94.

16. Dull, T., Zufferey, R., Kelly, M., Mandel, R.J., Nguyen, M., Trono, D., and Naldini, L., 1998. A Third-Generation Lentivirus Vector with a Conditional Packaging System. *Journal of Virology* 72:8463–8471.

17. Durai, S., Mani, M., Kandavelou, K., Wu, J., Porteus, M. H., and Chandrasegaran, S. 2005. Zinc finger nucleases: custom-designed molecular scissors for genome engineering of plant and mammalian cells. *Nucleic Acids Res.* 33:5978–5990.

18. Ebert, A. D., Beres, A. J., Barber, A. E., and Svendsen, C. N. 2007. Human neural progenitor cells over-expressing

IGF-1 protect dopamine neurons and restore function in a rat model of Parkinson's disease. *Exp. Neurol.* 209:213–223.

19. Ehrlich, M., Sarafyan, L. P., and Myers, D. J. 1976. Interaction of microbial DNA with cultured mammalian cells. Binding of the donor DNA to the cell surface. *Biochim. Biophys. Acta.* 454:397–409.

20. Emery D.W., Nishino T., Murata K., Fragkos M., and Stamatoyannopoulos G. 2002. Hematopoietic stem cell gene therapy. *Int. J. Hematol.* 75:228–236.

21. Felgner, J. H., Kumar, R., Sridhar, C. N., Wheeler, C. J., Tsai, Y. J., Border, R., Ramsey, P., Martin, M., and Felgner, P. L. 1994. Enhanced gene delivery and mechanism studies with a novel series of cationic lipid formulations. *J. Biol. Chem.* 269:2550–2561.http://www.ncbi.nlm.nih.gov/pubmed/8300583/

22. Felgner, P. L., Gadek, T. R., Holm, M., Roman, R., Chan, H. W, Wenz, M., Northrop, J. P., Ringold, G. M., and Danielsen, M. 1987. Lipofection: a highly efficient, lipid-mediated DNA-transfection procedure. *Proc. Natl. Acad. Sci. USA* 84:7413–7417.

23. Galimi F. and Verma I.M. 2002. Opportunities for the use of lentiviral vectors in Human Gene Therapy. *Current Topics in Microbiology and Immunology* 261:245–253. From the Volume on *Lentiviral Vectors* edited by Trono, D.

24. Ghosh-Choudhury, G., Haj-Ahmad, Y., Brinkley, P., Rudy, J., and Graham, F. L. 1986. Human adenovirus cloning vectors based on infectious bacterial plasmids. *Gene* 50:161–171

25. Graham, F. L., Smiley, J., Russell, W. C., and Nairn, R. 1977. Characteristics of a human cell line transformed by DNA from human adenovirus type 5. *J. Gen. Virol.* 36:59–74.

26. Gugala, Z., Olmsted-Davis, E. A., Gannon, F. H., Lindsey, R. W. and Davis, A. R. 2003. Osteoinduction by ex vivo adenovirus-mediated BMP2 delivery is independent of cell type. *Gene Ther.* 10:1289–1296.

27. Hartman, Z. C., Appledorn, D. M., and Amalfitano, A. 2008. Adenovirus vector induced innate immune responses: impact upon efficacy and toxicity in gene therapy and vaccine applications. *Virus Res.* 132:1–14.

28. Howe, S. J., Mansour, M. R., Schwarzwaelder, K., Bartholomae, C., Hubank, M., Kempski, H., Brugman, M. H., Pike-Overzet, K., Chatters, S. J., de Ridder, D., Gilmour, K. C., Adams, S., Thornhill, S. I., Parsley, K. L., Staal, F. J., Gale, R. E., Linch, D. C., Bayford, J., Brown, L., Quaye, M., Kinnon, C., Ancliff, P., Webb, D. K., Schmidt, M., von Kalle, C., Gaspar, H. B., and Thrasher, A. J. 2008. Insertional mutagenesis combined with acquired somatic mutations causes leukemogenesis following gene therapy of SCID-X1 patients. *J. Clin. Invest.* 118:3143–3150

29. Hsu, W. K., Sugiyama, O., Park, S. H., Conduah, A., Feeley, B. T., Liu, N. Q., Krenek, L., Virk, M. S., An, D. S., Chen, I. S., and Lieberman, J. R. 2007. Lentiviral-mediated BMP-2 gene transfer enhances healing of segmental femoral defects in rats. *Bone* 40:931–938.

30. Izsvák, Z., Chuah, M. K., Vandendriessche, T., and Ivics, Z. 2009. Efficient stable gene transfer into human cells by the Sleeping Beauty transposon vectors. *Methods* Epub. Jul 15 2009.

31. Järås, M., Brun, A. C., Karlsson, S., and Fan, X. Adenoviral vectors for transient gene expression in human primitive hematopoietic cells: applications and prospects. *Exp Hematol.* 35:343–349.

32. Kafri, T., Blomer, U., Peterson, D.A., Gage, F.H., and Verma I.,M., 1997. Sustained expression of genes delivered directly into liver and muscle by lentiviral vectors. *Nature Genetics* 17:314–317

33. Kawabata, K., Sakurai, F., Koizumi, N., Hayakawa, T., and Mizuguchi, H. 2006. Adenovirus vector-mediated gene transfer into stem cells. *Mol. Pharm.* 3:95–103.

34. Kingston, R. E., Chen, C. A., and Okayama, H. 2001. Calcium phosphate transfection. *Curr. Protoc. Immunol.* Ch10, Unit 10.13.

35. Kim, D. and Rossi, J. 2008. RNAi mechanisms and applications. *Biotechniques* 44:613–616.

36. Kim, E. Y., Hong, Y. B., Lai, Z., Cho, Y. H., Brady, R. O., and Jung SC. 2005. Long-term expression of the human glucocerebrosidase gene in vivo after transplantation of bone-marrow-derived cells transformed with a lentivirus vector. *J. Gene. Med.* 7:878–887

37. Korswagen, H. C. and Clevers, H. C. 1999. Activation and repression of wingless/WnT target genes by the TCF/LEF-1 family of transcription factors. *Cold Spring Harb. Symp. Quant Biol.* 64:141–147.

38. Loebinger M. R., Eddaoudi A., Davies D., and Janes S. M. 2009. Mesenchymal stem cell delivery of TRAIL can eliminate metastatic cancer. *Cancer Res.* 69:4134–4142.

39. Logan, A. C., Lutzko, C., and Kohn, D. B. 2002. Advances in lentiviral vector design for gene-modification of hematopoietic stem cells. *Curr. Opin. Biotechnol.* 13:429–436.

40. Lombardo, A., Genovese, P., Beausejour, C. M., Colleoni, S., Lee, Y. L., Kim, K. A., Ando, D., Urnov, F. D., Galli, C., Gregory, P. D., Holmes, M. C., and Naldini, L. 2007. Gene editing in human stem cells using zinc finger nucleases and integrase-defective lentiviral vector delivery. *Nat. Biotechnol.* 25:1298–1306.

41. Lu, L., Zhao, C., Liu, Y., Sun, X., Duan, C., Ji, M., Zhao, H., Xu, Q., and Yang, H. 2005. Therapeutic benefit of TH-engineered mesenchymal stem cells for Parkinson's disease. *Brain Res. Brain Res. Protoc.* 15:46–51.

42. Marchenko, S. and Flanagan, L. 2007. Transfecting human neural stem cells with the Amaxa Nucleofector. *J. Vis. Exp.* 6:240.

43. Marini, F. C., Shayakhmetov, D., Gharwan, H., Lieber, A., and Andreeff, M. 2002. Advances in gene transfer into haematopoietic stem cells by adenoviral vectors. *Expert. Opin. Biol. Ther.* 2:847–856.

44. Mei, S.J.H, McCarter, S. D., Deng, Y., Parker, C. H., Liles, W.C., and Stewart, D. J. 2007. Prevention of LPS-Induced Acute Lung Injury in Mice by Mesenchymal Stem Cells Overexpressing Angiopoietin 1. *PLoS Med.* 4:e269.

45. Naqvi, A. R., Islam, M. N., Choudhury, N. R., and Haq, Q. M. 2009. The fascinating world of RNA interference. *Int. J. Biol. Sci.* 5:97–117.

46. Meyerrose, T. E., Roberts, M., Ohlemiller, K. K., Vogler, C. A., Wirthlin, L., Nolta, J. A., and Sands, M. S. 2008. Lentiviral-transduced human mesenchymal stem cells persistently express therapeutic levels of enzyme in a xenotransplantation model of human disease. *Stem Cells* 26:1713–22.

47. Pagnotto, M. R., Wang, Z., Karpie, J. C., Ferretti, M., Xiao, X., and Chu, C. R. 2007. Adeno-associated viral gene transfer of transforming growth factor-beta1 to human mesenchymal stem cells improves cartilage repair. *Gene Ther.* 14:804–813.

48. Papapetrou, E. P., Zoumbos, N. C., and Athanassiadou, A. 2005. Genetic modification of hematopoietic stem cells with nonviral systems: past progress and future prospects. *Gene Ther.* S1:S118–30.

49. Peister, A., Mellad, J.A., Wang, M., Tucker, H.A., and Prockop, D.J. 2004. Stable transfection of MS cells by electroporation. *Gene Ther.* 11:224–228.

50. Perumbeti, A., Higashimoto, T., Urbinati, F., Franco, R., Meiselman, H. J., Witte, D., and Malik, P. 2009. A novel human gamma-globin gene vector for genetic correction of sickle cell anemia in a humanized sickle mouse model: critical determinants for successful correction. *Blood* 114:1174–1185.

51. Porteus, M. H. and Carroll, D. 2005. Gene targeting using zinc finger nucleases. *Nat. Biotechnol.* 23:967–973.

52. Qing, K., Mah, C., Hansen, J., Zhou, S., Dwarki, V., and Srivastava, A. 1999. Human fibroblast growth factor receptor 1 is a co-receptor for infection by adeno-associated virus 2. *Nat Med.* 5:71–77.

53. Rabin, N., Kyriakou, C., Coulton, L., Gallagher, O. M., Buckle, C., Benjamin, R., Singh, N., Glassford, J., Otsuki, T., Nathwani, A. C., Croucher, P. I., and Yong, K. L. 2007. A new xenograft model of myeloma bone disease demonstrating the efficacy of human mesenchymal stem cells expressing osteoprotegerin by lentiviral gene transfer. *Leukemia* 21:2181–2191.

54. Raimondo, S., Penna, C. Pagliaro, P., and Geuna, S. 2006. Morphological characterization of GFP stably transfected

adult mesenchymal bone marrow stem cells. *J. Anat.* 208:3–12.

55. Sadelain M., Lisowski L., Samakoglu S., Rivella S., May C., and Riviere I. 2005. Progress toward the genetic treatment of the beta-thalassemias. *Ann. NY Acad. Sci.* 1054:78–91.

56. Sakurai, H., Kawabata, K., Sakurai, F., Nakagawa, S., and Mizuguchi H. 2008. Innate immune response induced by gene delivery vectors. *Int. J. Pharm.* 354:9–15.

57. Schenborn, E. T. and Goiffon, V. 2000. Calcium phosphate transfection of mammalian cultured cells. *Methods Mol. Biol.* 130:135–145.

58. Scherr, M. and Eder, M. 2002. Gene transfer into hemato-poietic stem cells using lentiviral vectors. *Curr. Gene Ther.* 2:45–55.

59. Scherr, M. and Eder, M. 2005. Modulation of gene expression by siRNA in hematopoietic cells. *Curr. Opin. Drug Discov. Devel.* 8:262–269.

60. Schmidt, M., Glimm, H., Wissler, M., Hoffmann, G., Olsson, K., Sellers, S., Carbonaro, D., Tisdale, J. F., Leurs, C., Hanenberg, H., Dunbar, C. E., Kiem, H. P., Karlsson, S., Kohn, D. B., Williams, D., and Von Kalle, C. 2003. Efficient characterization of retro-, lenti-, and foamyvector-transduced cell populations by high-accuracy insertion site sequencing. *Ann. NY Acad. Sci.* 996:112–121.

61. Schroder, A. R., Shinn, P., Chen, H., Berry, C., Ecker, J. R., and Bushman, F. 2002. HIV-1 integration in the human genome favors active genes and local hotspots. *Cell* 110:521–529.

62. Shi, P. A., Hematti, P., Von Kalle, C., and Dunbar, C. E. 2002. Genetic marking as an approach to studying in vivo hematopoiesis: progress in the non-human primate model. *Oncogene* 21:3274–3283.

63. Srivastava A. 2005. Hematopoietic stem cell transduction by recombinant adeno-associated virus vectors: problems and solutions. *Hum. Gene. Ther.* 16:792–798.

64. Sugiyama, O., An, D. S., Kung, S. P., Feeley, B. T., Gamradt, S., Liu, N. Q., Chen, I. S., and Lieberman, J. R. 2005. Lentivirus-mediated gene transfer induces long-term transgene expression of BMP-2 in vitro and new bone formation in vivo. *Mol Ther.* 11:390–398

65. Summerford, C. and Samulski, R. J. 1998. Membrane-associated heparan sulfate proteoglycan is a receptor for adeno-associated virus type 2 virions. *J. Virol.* 72:1438–1445.

66. Summerford, C., Bartlett, J. S., and Samulski, R. J. 1999. AlphaVbeta5 integrin: a co-receptor for adeno-associated virus type 2 infection. *Nat Med.* 5:78–82.

67. Tolar, J., Nauta, A. J., Osborn, M. J., Panoskaltsis Mortari, A., McElmurry, R. T., Bell, S., Xia, L., Zhou, N., Riddle, M., Schroeder, T. M., Westendorf, J. J., McIvor, R. S., Hogendoorn, P. C., Szuhai, K., Oseth, L., Hirsch, B., Yant, S. R., Kay, M. A., Peister, A., Prockop, D. J., Fibbe, W. E., and Blazar, B. R. 2007. Sarcoma derived from cultured mesenchymal stem cells. *Stem Cells* 25:371–379.

68. Tsuda, H., Wada, T., Ito, Y., Uchida, H., Dehari, H., Nakamura, K., Sasaki, K., Kobune, M., Yamashita, T., and Hamada, H. 2003. Efficient BMP2 gene transfer and bone formation of mesenchymal stem cells by a fiber-mutant adenoviral vector. *Mol. Ther.* 7:354–365

69. Van Damme, A., Vanden Driessche, T., Collen, D., and Ch-uah, M. K. 2002. Bone marrow stromal cells as targets for gene therapy. *Curr. Gene Ther.* 2:195–209.

70. Vanhecke, D. and Janitz, M. 2005. Functional genomics using high-throughput RNA interference. *Drug Discov. Today* 10:205–212.

71. Wang, J., Ye, F., Cheng, L., Shi, Y., Bao, J., Sun, H., Wang, W., Zhang, P., and Bu, H. 2009. Osteogenic differentiation of mesenchymal stem cells promoted by overexpression of connective tissue growth factor. *J. Zhejiang. Univ. Sci B.* 10:355–367.

72. Wang, X., Cade, R., and Sun, Z. 2005. Expression of human eNOS in cardiac and endothelial cells. *Methods Mol. Med.* 112:91–107.

73. Wang, Z., Ma, H. I., Li, J., Sun, L., Zhang, J., and Xiao, X. 2003. Rapid and highly efficient transduction by double-stranded adeno-associated virus vectors in vitro and in vivo. *Gene Ther.* 10:2105–2111.

74. Weaver J. C. 1995a. Electroporation theory. Concepts and mechanisms. *Methods Mol. Biol.* 48:3–28.

75. Weaver J. C. 1995b. Electroporation theory. Concepts and mechanisms. *Methods Mol. Biol.* 47:1–26.

76. Weaver, J. C. 1993. Electroporation: a general phenomenon for manipulating cells and tissues. *J. Cell Biochem.* 51:426–435.

77. Wu, S., Sasaki, A., Yoshimoto, R., Kawahara, Y., Manabe, T., Kataoka, K., Asashima, M., and Yuge, L. 2008. Neural stem cells improve learning and memory in rats with Alzheimer's disease. *Pathobiology.* 75:186–194.

78. Wu, J., Kandavelou, K., and Chandrasegaran, S. 2007. Custom-designed zinc finger nucleases: what is next? *Cell Mol. LIFe Sci.* 64:2933–2944.

79. Yan, J., Tang, T., Li, F., Zhou, W, Liu, J., Tan, Z., Zheng, W., Yang, Y., Zhou, X., and Hu, J. 2009. Experimental Study of the Effects of Marrow Mesenchymal Stem Cells Transfected with Hypoxia-Inducible Factor-1α Gene. *J. Biomed. Biotechnol.* Epub. Jul 1 2009:128627.

80. Zhang, X. Y., La Russa, V. F., Bao, L., Kolls, J., Schwarzenberger, P., and Reiser, J. 2002. Lentiviral vectors for sustained transgene expression in human bone marrow-derived stromal cells. *Mol. Ther.* 5:555–565

81. Zou, G. M. and Yoder, M. C. 2005. Application of RNA interference to study stem cell function: current status and future perspectives. *Biol. Cell.* 97:211–219.

Chapter 8

Stem Cells for Neuroreplacement Therapies

Kiminobu Sugaya
Burnet School of Biomedical Science, College of Medicine, University of Central Florida, Orlando, Florida, USA

For correspondence:
Sugaya Kiminobu, PhD
Burnet School of Biomedical Science
College of Medicine
University of Central Florida
4000 Central Florida Blvd
BMS room 223, 220
Orlando FL 32816-2364

E-mail: ksugaya@mail.ucf.edu
Phone: (407) 823-1524
Fax: (407) 823-1524

Recent advances in stem cell technology are expanding our ability to replace a variety of cells throughout the body. In the past, neurological diseases caused by the degeneration of neuronal cells were considered incurable because of a long-held truism: neurons do not regenerate during adulthood. However, neural stem (NS) cells, which commit to become neurons and glial cells, have been found in adult mammalian brains, and now we have much evidence that the adult brain is indeed capable of regenerating neurons. Based on this new concept, researchers have shown NS cell transplantation to the brain can produce neural cells and recover function even in the aged brain. These results may promise a bright future for clinical applications of stem cell strategies in neurological diseases. Here, we describe different types of stem cells and discuss possible use of them in neuroreplacement therapies.

1. Different Types of Stem Cells

Stem cells are often defined by their ability to self-renew and potential to become a variety of cells. While pluripotent stem cells, like embryonic stem (ES) cells, can develop into any of the three major tissue types: endoderm (interior gut lining), mesoderm (muscle, bone, blood), or ectoderm (epidermal tissues and nervous system), the ability of multipotent stem cells found in adults to become a variety of cells is limited by their individual characteristics, which includes their partial commitment to a specific cell lineage. Each adult stem cell contains distinctive information that would allow it to become a discreet type of cell in a tissue-specific environment. That is, these cells are partially committed to become a certain type of stem cell in a tissue-specific manner. For example, hematopoietic stem cells isolated from bone marrow differentiate into a variety of blood cells but not into neural cells, and NS cells isolated from the brain differentiate into neurons and glia but not into blood cells under normal conditions.

The variability of stem cells also depends on the developmental stage of the source from which the stem cells are isolated. The commitment of stem cells to differentiate into particular cell types is higher in adult stem cells than in fetal or embryonic stem cells; additionally, the rate at which fetal or embryonic stem cells proliferate is faster than that of adult stem cells, which may limit the value of adult stem cells in clinical applications. There are ethical and immunological issues associated with embryonic or fetal stem cells, as well. Furthermore, ES cells are known to have a tendency to form tumors since they do not have strict instructions for the differentiation.

The tissue specificity of stem cells seems highly related to DNA methylation. Cytosine methylation of mammalian DNA is essential for the proper epigenetic regulation of gene expression and the maintenance of genomic integrity (Riggs 1989). Proliferating cells have an average of 15% more methylated cytosines than nondividing cells (Schwarz, Alexander, et al. 1999). 5-Azacytidine, a demethylation agent, enhances the maturation of neurons derived from endothelial growth factor–generated NS cells (Schinstine and Iacovitti 1997), indicating that DNA methylation is a mechanism that regulates the differentiation of stem cells. During development, stem cells existing in each tissue have achieved a well-accepted lineage commitment for tissue-specific differentiation. This allows them to generate tissue-specific cells only under normal conditions. Theoretically, lineage-specific differentiation of stem cells depends on the activation of specific transcription. Since DNA methylation regulates gene transcription, epigenetic regulation of DNA methylation patterns may be comparable to modifying the cell fate decisions of the cells.

Immunorejection could be prevented by use of ES cells from individual clones. However, current cloning technology by nuclear transfer is an inefficient process in which most clones die before birth, and survivors often display growth abnormalities (Humpherys, Eggan, et al. 2001); developmental anomalies of cloned embryos could be due to incomplete epigenetic reprogramming of donor genomic DNA methylation (Kang, Koo, et al. 2001). The unreliability of cloned tissue should be considered when ES cells are used as transplantation materials.

2. Neurogenesis in the Adult Brain

The discovery of multipotent NS cells in the adult brain (Alvarez-Buylla and Kirn 1997; Gould, Reeves, et al. 1999) has brought revolutionary changes in the theory of neurogenesis, which currently posits that regeneration of neurons can occur throughout life. This is particularly prominent in the subventricular zone (SVZ) and dentate gyrus of the hippocampal formation. Since newly generated neurons from NS cells are functional, adult neurogenesis may be a compensation mechanism for brain damage (Van Praag, Schinder, et al. 2002). Endogenous stem cells proliferate, migrate, and regenerate damaged neurons after ischemic neuronal injury (Nakatomi, Kuriu, et al. 2002). Olfactory neurons located in the olfactory epithelium are continuously replaced by NS cells migrating from the SVZ. In adults, although neurogenesis of olfactory neurons after lesioning was less efficient than in young mice, NS cells still retained their ability to provide functional recovery (Ducray, Bondier, et al. 2002). Patients having neurodegenerative diseases, including Alzheimer's and Parkinson's diseases, are know to have a diminished sense of smell. This may be related to the diminished neurogenesis in the olfactory system because of the high demand of stem cells in the different area of neuronal system in these patients.

The fact that the adult brain sill possesses the ability to guide and regulate stem cell biology stimulates us to investigate stem cell therapy for neurodegenerative diseases.

3. Effect of Aging on the Stem Cell Population in the Adult Brain

Although stem cell activity exists throughout life, the reported decline in neurogenesis as a result of aging may be related to a decreasing proliferation of stem cells, as mentioned above (Kuhn, Dickinson-Anson, et al. 1996). This reduction of stem cell population in the aged brain may partially explain why aging is the highest risk factor for neurodegenerative diseases and may be a rationale for stem cell augmentation therapies, including stem cell transplantation, for these diseases. While endogenous neuroregeneration in the adult brain may be minimal, and it is not clear whether neurogenesis is essential for normal cognitive function during aging, a defect or decreasing effectiveness of neuroregeneration might significantly decline normal brain function, especially if the endogenous stem cell system is affected by the diseases.

4. Origin and Characteristics of Neural Stem Cells

Neural stem cells, capable of spontaneously differentiating into neurons and glia, may be the most promising candidate for neuroreplacement therapies. Neural stem cells have been isolated from the fetal and adult mammalian (Doetsch, Caille, et al. 1999; Johansson, Svensson, et al. 1999) and human (Johansson, Momma, et al. 1999) central nervous system (CNS) and propagated in vitro in a variety of culture systems (Svendsen, Ter Borg, et al. 1998).

Immortalized fetal mouse NS cells have been studied for years (Sinden, Stroemer, et al. 2000). However, current interest is focused on the ability to proliferate populations of human NS cells as an in vitro source of tissue for brain repair since successful neuroreplacement by stem cell transplantation is dependent on these capacities (i.e., the ability of these cells not only to migrate but also to integrate into the host CNS; the source and culture conditions of stem cells are crucial factors). In many cases, NS cells are isolated by their characteristic spheroid formation under the influence of the basic fibroblast growth factor (bFGF) and the epidermal growth factor (EGF). Although ideal positive selection markers have not yet been discovered, phenotypical CD133 (+), CD34(-), CD45(-) cells from human fetal brain tissue seem to possess the ability to differentiate into neural cells (Uchida, Buck, et al. 2000).

The inability to grow NS cells in vitro in the absence of complex and undefined biological fluids (i.e., serums) had long been a major obstacle in understanding the biology of these cells. However, it is now possible to maintain and expand NS cells in serum-free media containing bFGF and EGF (Fricker, Carpenter, et al. 1999; Brannen and Sugaya 2000), which help to prepare transplantable material. Nonetheless, there

213

have been reports showing that hNS cells express almost nondetectable telomerase after certain passages and that these human cells also have significantly shorter telomeres than their rodent counterparts (Ostenfeld, Caldwell, et al. 2000). Some researchers have been able to continuously expand these cells in vitro (Brannen and Sugaya 2000). After long-term culturing using a serum-free unsupplemented media condition, human NS (hNS) cells differentiated into βIII-tubulin-, glial fibrillary acidic protein (GFAP)-, and O4- immunopositive cells, markers for neurons, astrocytes, and oligodendrocytes, respectively (Brannen and Sugaya 2000), suggesting that the genesis of neurons or astrocytes can take place without the addition of exogenous differentiation factors. Thus, it appears that hNS cells are capable of producing the endogenous factors necessary for their own differentiation and survival, which could help hNS cells to survive in the aged animal model. This ability to expand multipotent hNS cells in vitro under defined conditions offers a well-characterized and efficient source of transplantable cell materials.

5. Applications of Neural Stem Cells

The successful transplantation of hNS cells into aged rats with subsequent improvement of cognitive function reported by Qu et al. (Qu, Brannen, et al. 2001) reinforces the potential feasibility of hNS cell transplantation neuro-replacement therapy. In their study, hNS cells, expanded without differentiation under the influence of mitogenic factors in supplemented serum-free media (Qu, Brannen, et al. 2001) and labeled by the incorporation of bromodeoxyuridine (BrdU) into the nucleus DNA, were injected into the lateral ventricle of mature (six-month-old) and aged (twenty-four-month-old) rats. Cognitive function of the animals was assessed by the Morris water maze both before and four weeks after the transplantation of hNS cells. Before hNS cell transplantation, some aged animals (aged memory-unimpaired animals) cognitively functioned in the range of mature animals, while others (aged memory-impaired animals) functioned entirely below the cognitive range of the mature animals. After hNS cell transplantation, most aged animals had cognitive function in the range of the mature animals. Strikingly, one of the aged memory-impaired animals showed dramatic improvement in behavior, functioning even better than the mature animals. Statistical analysis showed that cognitive function was significantly improved in both mature and aged memory-impaired animals. These behavioral results show the beneficial effects of hNS cell transplantation into the host brain in most animals tested.

After the second water maze task, postmortem brains were further analyzed by immunohistochemistry for βIII-tubulin and GFAP, markers for neurons and astrocytes, respectively. There was no sign of ventricular distortion and no evidence of tumor formation, and further, no strong host anti-graft immunoreactivity was observed. Intensely and extensively stained with βIII-tubulin, neurons with BrdU-positive nuclei were found in bilateral cingulate and parietal cortexes and in the hippocampus. The βIII-tubulin-positive neurons found in the cerebral cortex were typified by a dendrite pointing to the edge of the cortex. In the hippocampus, donor-derived neurons exhibited multiple morphologies varying in cellular size and shape and one or more processes and branching.

Thus, hNS cell may be the most promising candidate for neuroreplacement therapy; however, ethical issues, difficulty accessing enough material, and the risk of immunological rejection limit their value.

215

6. Origin and Characteristics of Embryonic Stem Cells

Some researchers are trying to find a transplantable cell source in ES cells isolated from the inner cell mass of human embryos or blastocytes (Shamblott, Axelman, et al. 2001). Embryonic stem cells proliferate extensively and theoretically can differentiate into any type of somatic cells. These characteristics make ES cell strong potential candidates for cell therapies. However, we must develop methods of enriching the cells of interest because ES cells do not have the information necessary to become specific types of cells that may be needed in individual cases. For example, ES cells are not committed to become neural cells as NS cells are, and without that commitment, ES cells might grow into unwanted cells, possibly even forming tumors when they are transplanted into the brain. Although theoretically ES cells could be modified into cells possessing the same genetic material as the patient by somatic nuclei transfer and cloning (Munsie, Michalska, et al. 2000) to eliminate immunological problem, human cloning has not been successful so far. Other than the difficulty of human cloning, we have a barrier to overcome before developing an autologous cell therapy using ES cells: that is, tissue-specific epigenetic modifications. Cloning by nuclear transfer is an inefficient process in which most clones die before birth and survivors often display growth abnormalities (Rideout, Eggan, et al. 2001). This may be due to the tissue-specific DNA methylation pattern from somatic nuclei used in cloning (Humpherys, Eggan, et al. 2001). Although Munsie et al. reported that pluripotent stem cells can be derived from reprogrammed nuclei of terminally differentiated adult somatic cells (Munsie, Michalska, et al. 2000), cloned ES cells may also be modified by tissue-specific epigenesis that has occurred in the somatic cells, and may not fully function as neural cells. Thus, cloned ES cells may also receive

216

tissue-specific epigenetic modifications and may not fully function as neural cells.

7. Applications of Embryonic Stem Cells

Mackay's group reported that a highly enriched population of midbrain NS cells can be derived from mouse ES cells (Kim, Auerbach, et al. 2002). They reported that dopamine neurons generated by these stem cells show the electro-physiological and behavioral properties expected of neurons from the midbrain, encouraging us to consider the use of ES cells in cell replacement therapy for Parkinson's disease. These authors focused on the fact that midbrain precursor cells express *nuclear receptor related-1 (Nurr1)*, which is a transcription factor with a role in the differentiation of midbrain precursors into dopamine neuron. Further, they established a stable *Nurr1* expression ES cell line using cytomegalovirus plasmid and processed by the five-stage method (Lee, Lumelsky, et al. 2000) that leads to the efficient differentiation of ES cells into neurons. They found that the majority of ES cells differentiated into tyrosine hydroxylase-positive cells in vitro. Then they grafted these terminally differentiated ES cells into the rat striatum after 6-hydroxy dopamine lesioning. The animals that received grafts of the ES cell–derived dopaminergic neurons showed functional recovery in amphetamine-induced rotation behavior and also showed that the transplanted cells responded to electrical stimuli, similar to host striatal neurons.

Although their results are promising in neuroreplacement strategies using ES cells for treatment of neurological diseases, long-term safety and efficacy issues need to be addressed. For example, ES cells are not committed to become

217

neural cells as NS cells are, and without that commitment, unwanted cells, like tumors, may also be developed along with midbrain dopamine neurons.

8. Origin and Characteristics of Mesenchymal Stem Cells

Bone marrow contains stem cells used not only for hematopoiesis but also for production of a variety of nonhematopoietic tissues. A subset of stromal cells in bone marrow, which has been referred to as mesenchymal stem (MS) cells, is capable of producing multiple mesenchymal cell lineages, including bone, cartilage, fat, tendons, and other connective tissues (Pereira, Halford, et al. 1995; Prockop 1997; Majumdar, Thiede, et al. 1998; Pittenger, Mackay, et al. 1999). Several reports show that human MS (hMS) cells also have the ability to differentiate into a diverse family of cell types that may be unrelated to their phenotypical embryonic origin, including muscle and heptocytes (Ferrari, Cusella-De Angelis, et al. 1998; Makino, Fukuda, et al. 1999; Petersen, Bowen, et al. 1999; Liechty, MacKenzie, et al. 2000; Imasawa, Utsunomiya, et al. 2001; Mackenzie and Flake 2001). Although the potential therapeutic use of hMS cells in the central nervous system has been discussed (Prockop, Azizi, et al. 2000; Bianco, Riminucci, et al. 2001), and several in vivo transplantation studies showed neural and glial differentiation of hMS cells (Kopen, Prockop, et al. 1999; Schwarz, Alexander, et al. 1999; Chen, Li, et al. 2000; Chopp, Zhang, et al. 2000; Li, Chopp, et al. 2000), technologies to induce neural lineage from hMS cells are not fully established. Adult stem cells continue to possess some multipotency, but cell types produced from adult stem cells are limited by their tissue-specific character. Alterations are necessary to overcome this barrier of stem cell lineage.

However, the regulation mechanisms of tissue-specific stem cell fate decisions remain unclear. Thus, to differentiate MS cells into neural cells, alteration of their epigenetic information before transplantation may be necessary. Other issues associated with MS cells are the inability to grow the cells in vitro without serum and the absence of positive selection markers for definitive isolation of these cells. Thus, neuroreplacement therapy by MS cell transplantation must clear some hurdles before it can be considered for clinical use.

9. Applications of Mesenchymal Stem Cells

Verfaillie's group reported that they identified multipotent progenitor cells that co-purify with MS cells in adult bone marrow (Jiang, Jahagirdar, et al. 2002). They claimed these cells contribute to most, if not all, somatic cell types, and this subpopulation of hMS cells may be able to differentiate into neural cells. However, other researchers have not yet successfully reproduced the experiment. Under basal media conditions, hNS cells spontaneously differentiated into neural cells without additional factors; however, hMS cells never spontaneously differentiated into neural cells. As described above, these results indicate that each adult stem cell contains specific information that would allow it to become a special type of cell: that is, cells are partially committed to differentiate in a tissue-specific manner.

Two different groups reported spontaneous fusion of stem cells (Terada, Hamazaki, et al. 2002; Ying, Nichols, et al. 2002). In these reports, the authors found that stem cells acquired phenotypes from other cells by fusion, which may occur when these stem cells directly touch other cells after transplantation. The fusion of somatic cells to embryonic stem cells prompts expression of the embryonic stem cell gene *Oct4*, which may be initiated by signaling within the host cell, similar to nuclear reprogramming following nuclear transfer. Expression of stem cell genes that regulate self-renewal and pluripotency, such as *Oct4*, is likely to play an integral role in cellular reprogramming. Earlier work has indicated that the expression of certain genes makes it possible to maintain ES cells in a pluripotent state. *Nanog* is a divergent homeodomain protein that expresses in pluripotent ES cells. *Nanog* is a major regulator of ES cells and overrides signals to differentiate when over-expressed and is capable of maintaining ES cell self-renewal independently of *LIF/STAT3* (Chambers, Colby, et al. 2003). The suppression of differentiation has been previously demonstrated with the over expression of other ES cell genes, including *Pem* (Fan, Melhem, et al. 1999) and *Rex1* (Eiges, Schuldiner, et al. 2001), although elevated levels of *Oct4* were insufficient to guard against ES cell differentiation. We took a different approach and proposed that developmental potency can be gained by changing the gene expression profile through the embryonic stem cell gene *Nanog*, without the need for cell fusion. Modified cells can be recommitted to develop along a neuronal lineage and serve as a means of treating neurodegenerative conditions.

The *Nanog* gene sequence can be segmented into seven distinct regions: the 5′ untranslated region (UTR), N-terminal domain, homeodomain, C1 domain, Cw domain, C2 domain, and the 3′ UTR. The 5′ region contains binding sites for embryonic stem cell genes *Oct4* and *Sox-2*, which are part of a transcriptional regulatory loop, as well as a *p53* binding site within the Nanog promoter region that facilitates ES cell differentiation (Lin, Chao, et al. 2005) and possibly the shift in replication timing that is observed with neural differentiation (Perry, Sauer, et al. 2004). The N-terminal region of *Nanog* has transcriptional activity (Pan and Pei 2003) and encodes for a sequence containing a *SMAD4* domain (Hart, Hartley, et al. 2004). The homeodomain portion is similar to the NK-2 and ANTP family of homeodomain transcription factors, but comparing 120 different homeodomain proteins using BLOcks SUbstitution Matrix (BLOSUM) and Point Accepted Mutation matrices suggests that Nanog represents a distinct protein family divergent from both the NK-2 and distal-less gene family. The C-terminal domain contains no apparent transactivation motifs, but has greater transactivation activity compared to the N-terminal and homeodomain (Pan and Pei 2003; Pan and Pei 2005). The C-terminal domain can be further subdivided into three regions: the portion immediately following the homeodomain region (C1), the subregion containing a unique repeated motif of tryptophan flanked with four polar-uncharged amino acids (Cw), and a more distal sequence (C2).

hMS cells over-expressing *Nanog* produced interesting cell types that formed clusters, while mock transfected cells receiving a plasmid vector containing green fluorescence protein (GFP) as a control maintained their fibroblast-like morphology. The majority of GFP-positive cell clusters consisted of smaller cells with reduced cytoplasm and higher rates of proliferation. The proliferative clusters tended to form either an adherent mass of cells (ES-like) or more spherical, non-adherent or loosely adherent, clumps somewhat resembling embryoid bodies (EB-like).

Following *Nanog* transfection, expression of both *Nanog* and *Oct4* were significantly elevated in hMS cells. Immunohistochemical staining of these cells using antibodies against ES cell markers *Nanog*, *Oct4*, *stage-specific embryonic antigen-3* (*SSEA3*), and *keratan sulfate-associated antigens TRA1-60* showed ES cell marker expression within the proliferative cell clusters. While the vast majority of naive hMS cells failed to stain for any ES cell markers, a low population of cells showed positive staining for transcription factor *Oct4*. Thus, over-expression of *Nanog* may results in the dedifferentiation of hMS cells and cause the expression of ES cell markers.

Dedifferentiated cells placed in co-culture consisted of neurons and glial cells derived from hNS cells separated by a semi-permeable membrane. This system allowed for the exchange of growth factors and eliminated the concern over cell fusion since it prevented direct cell contact between the treated cells and the underlying feeder cells. Dedifferentiated cell clusters adhered to the membrane surface and differentiation occurred as cells radiated outwards. Control MS cells adhered to the membrane surface but failed to differentiate into neurons and astrocytes. Differentiation pattern was tested by immunostaining against βIII-tubulin and GFAP, and neuronal and astrocytic marker, respectively. The un-transfected MS cells did not show positive staining for the neuronal marker βIII-tubulin, but approximately 2% of the cells did show weak expression of GFAP. This may represent a subpopulation of pluripotent MS cells that is capable of astrocyte differentiation. Modified cells formed spherical clusters with a similar appearance to neural stem cell and stained positive for both βIII tubulin and GFAP. With extended culture time, the spherical cluster adhered to the membrane surface and cells began radiating outwards. Differentiated cells formed a web-like network of neurons and astrocytes.

Nanog transfection produced proliferative cells with morphological and gene expression similarities to embryonic stem cells. This study demonstrates a novel method of dedifferentiating adult stem cells by expressing genes regulating pluripotency with the end goal of facilitating neural transdifferentiation without reliance on cell fusion or the enrichment of low frequency subpopulations of cells.

222

9. Future of Transplantable Cell Materials for Neuroreplacement

In November 2007, two independent research teams' studies, one by a University of Wisconsin–Madison group (Yu, Vodyanik, et al. 2007) and another by Kyoto University, Japan (Takahashi, Okita, et al. 2007) created iPS cells from adult human cells. With the same principle used earlier in mouse models, Yamanaka had successfully transformed human fibroblasts into pluripotent stem cells using the same four pivotal genes: *Oct-3/4*, *Sox2*, *Klf4*, and *c-MYC* with a retroviral system. Thomson and colleagues used *Oct4*, *Sox2*, *Nanog*, and *LIN28*, using a lentiviral system. *Oct3/4* and certain members of the Sox gene family (*Sox1*, *Sox2*, *Sox3*, and *Sox15*) have been identified as crucial transcriptional regulators involved in the induction process.

Oct-3/4, one of the families of octamer (Oct) transcription factors, plays a crucial role in maintaining pluripotency. The absence of *Oct-3/4* in *Oct-3/4*⁺ cells, such as blastomeres and embryonic stem cells, leads to spontaneous trophoblast differentiation. The presence of *Oct-3/4* thus gives rise to the pluripotency and differentiation potential of embryonic stem cells. Various other genes in the Oct family, including *Oct-3/4*'s close relatives, *Oct1* and *Oct6*, fail to elicit induction, thus demonstrating the exclusiveness of *Oct-3/4* to the induction process. The Sox family of genes is associated with maintaining pluripotency similar to *Oct-3/4*. In contrast to *Oct-3/4*, which is exclusively expressed in pluripotent stem cells, *Sox2* is associated with multipotent and unipotent stem cells. While *Sox2* was the initial gene used for induction of pluripotency (Brambrink, Foreman, et al. 2008; Yamanaka 2008), other genes in the Sox family have been found to work as well in the induction process. *Sox1* yields iPS cells with a similar efficiency as Sox2, and *Sox3*, *Sox15*, and *Sox18* also generate iPS cells, although with decreased efficiency. Additional genes, like *Nanog*, have been identified to increase the induction efficiency.

In embryonic stem cells, Nanog, along with *Oct-3/4* and *Sox2*, is necessary in promoting pluripotency. Thomson et al. has reported it is possible to generate iPS cells with *Nanog* as one of the factors. *Nanog* over-expression successfully produced iPS cells from MS cells from adult bone marrow (Sugaya, Alvarez, et al. 2006). Nanog-transfected MS cells had not only a more immature morphology, as detected by smaller, rounder, shape, but also expressed *Oct4*, *Sox2*, and other embryonic stem cell markers. Additionally, the transfected cells possessed the ability to form embryobody-like cell mass and differentiate into neural cells. Thus, additional Nanog expression in the adult stem cells may be enough to increase their potency, which is much more practical for the clinical applications since other technologies require multiple gene transfection to the cells. Autologous cell transplantation using human skin blast derived iSP cells with over-expression of *Oct4*, *Sox-2*, and *Nanog*, and hMS cell–derived iPS cells with over-expression of *Nanog* may be a promising future strategy for neuroreplacement therapy.

Chapter 9

Endoderm

Keiichi Katsumoto,[1,2] Rika Miki,[1,2] Kahoko Umeda,[1] Nobuaki Shiraki,[1] Shoen Kume[1,2]

[1]Department of Stem Cell Biology, Institute of Molecular Embryology and Genetics (IMEG), Kumamoto University, Honjo 2-2-1, Kumamoto 860-0811, Japan
[2]The Global COE Cell Fate Regulation Research and Education Unit, Kumamoto University, Honjo 2-2-1, Kumamoto 860-0811, Japan

For correspondence:
Shoen Kume, PhD
Department of Stem Cell Biology
Institute of Molecular Embryology and Genetics
Kumamoto University
Honjo 2-2-1, Kumamoto 860-0811, Japan

Email: skume@kumamoto-u.ac.jp
Fax: +81-96-373-6807

1. Introduction

Stem cells can be defined as having the ability to self-renew and to generate differentiated cells. They are defined to demonstrate the following abilities [39]: the potency to reconstitute tissue throughout the organism's life span (long-term reconstitution), and the potency to give rise to multiple differentiated cell types (multipotency). Tissue stem cells are normally quiescent but become proliferative upon injury. To maintain a pool of stem cells in adult tissues, several cells must undergo cell divisions without differentiation. Asymmetric cell division is well known as a mechanism for maintaining a stem cell pool, and their progeny will proliferate and differentiate to generate a given tissue [19, 22, 34].

In contrast, progenitor cells are defined as the progeny of stem cells. Early progenitor cells have a multi-lineage potential, but late progenitor cells have restricted differentiation potential so that some differentiate into only a single lineage. Late progenitor cells do not self-renew, but they have the ability for short-term tissue reconstitution. They are also called "transit amplifying cells" [22, 29].

To understand the biology of stem cells, the following questions should be addressed. Where do they come from? Where do tissue stem cells reside? How do they differentiate into specific cell types? How are the senescence, proliferation, and differentiation of stem cells controlled? What are the stem cell niche and key mobilizing signals?

During embryogenesis, cells are initially proliferative and pluripotent and then they gradually become restricted to different cell fates. Knowledge from developmental biology and insights into the properties of stem cells are keys to further understanding and successful manipulation. Here, we first focus on the endoderm-derived organ development, such as the stomach, liver, pancreas, and intestine, on their cell types, the origin of the progenitors, and signals known in the maintenance or differentiation of the progenitors. We then focus on the tissue stem cells, such as liver, pancreas, and intestine, on their identification and characterization.

2. Embryonic Development

2.1 Stomach

Cytodifferentiation of stomach epithelium initiates around 13.5 dpc. The whole human stomach consists of glandular epithelium. In contrast, the mouse stomach consists of a squamous epithelium in the first proximal one-third. The distal two-thirds of the stomach consists of columnar glandular epithelium.

2.1.1 Hedgehog signaling

Figure 1 shows a schematic drawing of the hedgehog signalings in the developing stomach in mice. *Sonic hedgehog (Shh)* is expressed at a high level in the forestomach, but at a lower level in the hindstomach. *Indian hedgehog (Ihh)* is expressed in the hindstomach. At E18.5, *Shh* expression expands to the hindstomach. *Shh* and *Ihh* are both expressed in the glandular hindstomach (figure 1).

Fibroblast growth factor (FGF) signaling from the surrounding mesenchyme promotes the expression of *Ihh* in the glandular hindstomach and inhibits the expression of *Shh* in the hindstomach (figure 1). *Fgf10-/-* mice lack *Ihh* and showed a posterior expansion of *Shh* (table 1). *Fgf10-/-* mice showed a posterior expansion of the squamous epithelium. *Shh-/-* mice showed a reduced squamous epithelium in the forestomach and an increased glandular growth. Taken together, *Shh* signaling has a role in the specification of the forestomach by inhibiting the formation of the gland, which results in the generation of the squamous epithelium (table 2). In contrast, *Ihh-/-* mice did not show significant abnormality in the stomach region. Since there is redundancy in the Hedgehog pathway, and both *Shh* and *Ihh* are expressed in the glandular stomach, the exact role of *Ihh* awaits further investigation [37].

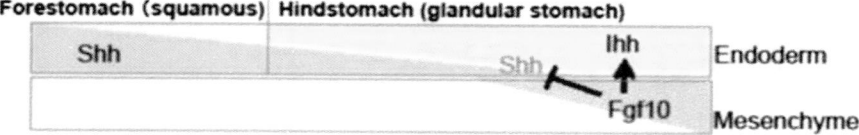

Table1 : Mutant mice phenotypes

Fgf10 -/-	Ihh : not expressed
	Shh : posterior expansion
	Posterior expansion of squamous epithelium (no gland)
Shh -/-	Gland formation : increased
Ihh -/-	No gross abnormalities
	(Gland formation is somewhat reduced)

Table2 : A high level of *Shh* expression in the forestomach inhibits gland formation.

	Shh expression level	Gland formation
forestomach	high	Inhibited squamous epithelium formation
hindstomach	low	Increased gland branching

Figure 1. A schematic drawing of the function of hedgehog signaling in the developing stomach in the mice.

2.2 Liver

2.2.1 Progenitors

The genes for albumin (*Alb*), transthyretin (*Ttr*) and α-fetoprotein (*Afp*) are among the earliest markers expressed during mammalian hepatic differentiation. These genes start to be expressed at the 7-somite stage in mice (8.25 dpc) and Alb/Afp double-positive hepatoblasts are bipotential progenitors that differentiate into hepatocytes and bile duct cells. Fate mapping studies showed that the pre-liver region exists in two separate lateral domains of the ventral endoderm and in the medial ventral endoderm at an early somite stage. These liver progenitor cells translocate to the anterior medial ventral region and are brought together during development [41].

2.2.2 FGF and BMP

FGF, secreted from the adjacent cardiac mesoderm, is reported to induce hepatic gene expression in the ventral foregut endoderm. The prehepatic endoderm receives a low concentration of FGF from the adjacent developing cardiac mesoderm, which permits liver differentiation through mitogen-activated protein kinase (MAPK). Later, the septum transversum develops between the liver primordium and cardiac mesoderm. This acts as a barrier against direct signaling to the liver primordium from the high concentration of FGF secreted from the cardiac mesoderm. This higher concentration of FGF induces lung development. The septum transversum secretes the bone morphogenetic protein 2 (BMP 2) and BMP 4, which are also important for generating liver progenitor cells [31, 41].

2.2.3 Transcription factors

Liver primordium is differentiated from the ventral endoderm, which express *SRY-box containing gene 17 (Sox17)*, *forkhead box A1 (Foxa1*, also known as *Hnf3a)*, *forkhead box A2 (Foxa2*, also known as *Hnf3β)*, *GATA binding protein 4 (Gata4)*, *Gata6*, and *HNF1 homeobox B (Hnf1b*, also known as *Tcf2)*. *Foxa1* and *Foxa2* have crucial roles in liver development. In *Foxa1* and *Foxa2* double-knockout mice, no liver bud is generated, and *Afp* cannot be detected in the foregut endoderm. Furthermore, in endoderm culture experiments of *Foxa1* and *Foxa2* double-deficient mice, *Ttr* and *Alb* cannot be induced, regardless of the presence of FGF2. Therefore, *Foxa1* and *Foxa2* expression establish the competence in the foregut endoderm to react to the hepatic-inducing signals. An early hepatoblast morphogenic event, from a cuboidal to a columnar shape generating pseudostratified epithelium, is controlled by the homeobox transcription factor gene *hematopoietically expressed homeobox (Hhex)*. The basal lamina is broken down and the cells move into the surrounding stroma and proliferate. These morphogenetic events are controlled by *one cut domain, family member 1 (Onecut1*, also known as *OC1* or *Hnf6)* and *one cut domain, family member 2 (Onecut2*, also known as *OC2)*. *Onecut1* and *Onecut2* regulate cell adhesion and migration-related gene expression patterns, such as *E-cadherin*, *thrombospondin 4 (Thbs4)* and *Secreted phosphoprotein 1 (Spp1*, also known as *osteopontin)*. Hepatoblast progenitor cells in the stroma receive signals from the endothelial cells to undergo maturation. Moreover, *prospero-related homeobox 1 (Prox1)* is involved in the delamination of hepatoblasts, and *T-box 3 (Tbx3)* has a role in mediating proliferation and is required for hepatoblast migration. The liver progenitor cells differentiate into ductal cells when Notch signaling is activated. *Onecut1* is the regulator of biliary development and the *hepatic nuclear* factor 4, alpha (Hnf4a) is expressed in parenchymal cells [20]. The cell lineages and molecules involved in liver development are summarized in figure 2.

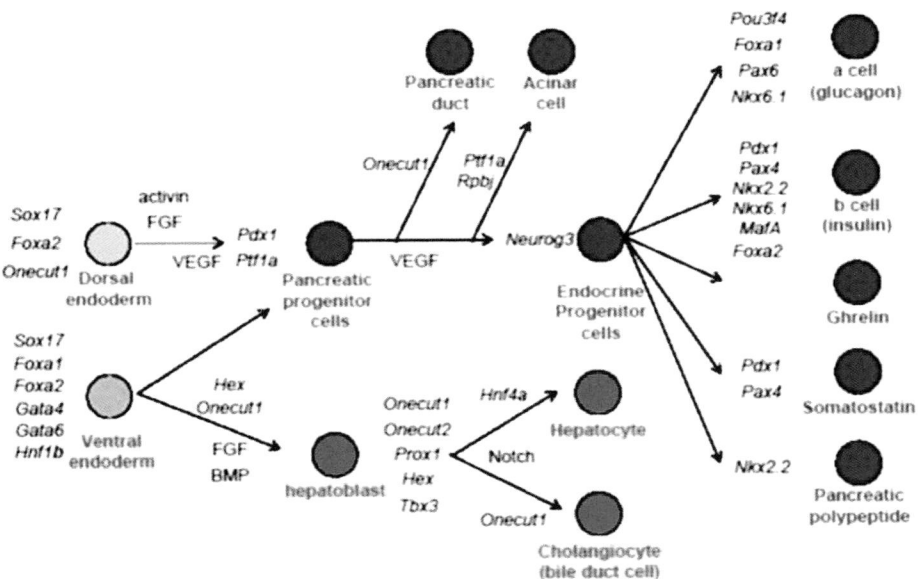

Figure 2. The molecules involved in the developing pancreas and liver.

2.3 Pancreas

The pancreas is composed of endocrine, exocrine, and ductular cells. The exocrine cells make up about 95–99% of the total pancreas. The endocrine cells, which compose the islets of Langerhans scattered in the exocrine tissue, make up about 1–5% of the total pancreas in the mouse. The islets of Langerhans are composed of α, β, δ, ε, and PP cell. These produce hormones such as glucagon, insulin, somatostatin, pancreatic polypeptide, and ghrelin. Each islet is made up from a central core of βcells surrounded by the other endocrine cells. Insulin down-regulates blood glucose level, whereas glucagon upregulates it. Ghrelin and pancreatic polypeptide are orexigenic hormones, and somatostatin controls the secretion of insulin, glucagons, and pancreatic polypeptide [4].

2.3.1 Pancreas progenitors

The pancreas starts to differentiate from the foregut/midgut junction of the endoderm at the 6–10 somite stage (8.5 dpc) in the mouse. The earliest pancreatic marker gene, *Pancreatic and duodenal homeobox1 (Pdx1)*, is expressed in the dorsal and ventral pancreatic bud and a portion of the stomach and duodenal endoderm. The dorsal pancreas is differentiated from the dorsal endoderm, which expresses *Sox17*, *Foxa2*, and *Onecut1*, and the ventral pancreas is generated from the ventral endoderm, which expresses *Sox17*, *Foxa1*, *Foxa2*, *Gata4*, *Gata6*, and *Hnf1b*. The dorsal pancreatic progenitor cells first appear near Hensen's node immediately after the completion of gastrulation, whereas the ventral pancreas progenitor cells begin to appear at the somite border at the 4-somite level near the vitelline vein at the 17 somite stage in the chick embryo [16, 23]. *Hnf1b* is required for the initial ventral pancreatic bud generation but not for dorsal pancreas bud generation. *Gata4* is required for *Pdx1* expression in the ventral endoderm. *Onecut1* is involved in *Pdx1* and *Neurogenin 3 (Neurog3)* expression in the pancreatic endoderm. *Onecut1* also has an important role in pancreatic duct development.

Lineage tracing experiments have shown that all pancreatic cells are derived from the *Pdx1*-expressing precursor cells. *Pdx1*-deficient mice do form a pancreatic bud, but development is arrested at an early somite stage. Although *Pdx1* plays a key role in pancreatic differentiation, *Pdx1* is not a determination factor because an ectopic pancreas is not

232

generated when *Pdx1* is overexpressed in the pre-stomach or pre-intestinal regions in the chick embryo [10]. In contrast, the ventral pancreas arises from the ventral foregut endoderm, where bipotential progenitor cells can differentiate into either liver or pancreas at around 8.5 dpc (8–10 somite stage). FGF signals from the cardiac mesoderm induce liver differentiation, while suppression of the FGF signaling in the ventral foregut endoderm permits development of the ventral pancreas. These observations suggest that the default state of the ventral foregut endoderm is for pancreatic development.

2.3.2 Activin and FGF2

Activin and FGF2 (also known as bFGF) are thought to be factors secreted from the notochord that direct dorsal pancreatic morphogenesis at around Hamburger-Hamilton developmental stage 12 (16 somite stage) in the chick embryo and maintain the expression of early pancreatic genes through the repression of *Shh* expression. Because the notochord cannot induce the posterior nonpancreatic endoderm to express pancreatic marker genes, notochord signals are considered to be permissive rather than instructive. Mice deficient for *activin receptor IIA* (*Acvr2a*, also known as *ActRIIa*) and *activin receptor IIB* (*Acvr2b*, also known as *ActRIIB*) express *Shh* in the prospective dorsal pancreas region and pancreatic development is disrupted.

2.3.3 Vascular endothelial growth factor A

Vascular endothelial growth factor A (Vegfa, also known as Vegf) is well known as a maturation signal for the dorsal pancreas and potentiates insulin expression within the pancreatic endoderm. The first signs of morphological change (budding from the endoderm) occur at the 22–25 somite stage (9.5 dpc) in the mouse. Mice deficient for the Vegfa receptor gene, *kinase insert domain protein receptor (Kdr,* also known as *Flk1),* exhibit disruption in the formation of blood vessels and the dorsal mesenchyme. In *Kdr*-deficient mice, *Pdx1* expression is normal at the 20–25 somite stage, but *pancreas specific transcription factor, 1a (Ptf1a)* expression is reduced in the dorsal but not in the ventral pancreas. Hence, aortal endothelial cells induce the expression of *Ptf1a* in the dorsal pancreas. An overlapping expression of *Pdx1* and *Ptf1a* defines the dorsal pancreas region in the *Xenopus* embryo. In the *Xenopus*, it is also reported that Wnt/β-catenin pathway have an important role in anterior-posterior regionalization within the endoderm, in a manner similar to the nervous system. A forced Wnt/β-catenin signaling in the anterior endoderm inhibits foregut development. In contrast, inhibition of Wnt/β-catenin signaling in the posterior endoderm permits ectopic pancreas and liver development to take place. It is also reported that retinoic acid (RA) or FGF signaling have a crucial role in anterior-posterior patterning within the endoderm.

2.3.4 Islet 1, N-cadherin, and FGF10

The dorsal mesenchyme surrounding the dorsal pancreas has a very significant role in development of the dorsal pancreas. Mice deficient for *ISL1 transcription factor, LIM/homeodomain (Isl1)* or *N-cadherin* display a lack of dorsal mesenchyme and abnormality in the dorsal pancreas. FGF10 is known to be secreted from the dorsal mesenchyme and promotes the accumulation of *Pdx1*-positive pancreatic progenitor cells.

2.3.5 Notch signaling pathways

One of the Notch signaling target genes, *hairy and en-hancer of split family* (*Hes*), represses the expression of *Neurog3*, an endocrine progenitor marker gene and *cy-clin-dependent kinase inhibitor 1C (P57)* (*Cdkn1c*), which regulates the cell cycle. All pancreatic endocrine cells are generated from *Neurog3*-positive cells. *Neurog3* null mice are completely lacking in pancreatic endocrine cells and show defects in acinar morphogenesis. Over-expression of *Neurog3* or of the intracellular form of Notch 3 (a repressor of Notch signaling) (Notch 3-ICD) leads to premature endocrine cell differentiation at the expense of the exocrine lineage. On the other hand, over-expression of Pdx1-Notch 1-ICD leads to reduced numbers of *Neurog3*-positive and endocrine cells, and diminished acinar differentiation. Mice deficient for *recombination signal binding protein for immunoglobulin kappa J region* (*Rbpj*), *delta-like 1 (Drosophila)* (*Dll1*), or *Hes1* exhibit in-creased numbers of *Neurog3*-positive cells, immature en-docrine differentiation, and a so-called plastic pancre-as. These results suggest that Notch signaling in the early phase of pancreatic development has an important role for maintaining the undifferentiated state of *Neurog3*-positive progenitor cells.

2.3.6 Other intrinsic molecules

It has been suggested that the numbers of *Pdx1*-positive progenitor cells between 8.5 dpc and 12.5 dpc deter-mines the size of the pancreas. Furthermore, in explant experiments, *Pdx1*-deficient epithelium cannot mature in the presence of wild type mesenchyme. Thus Pdx1 might provide competency to the pancreatic epithelium to re-spond to growth factors from the mesenchyme. From 9.5

dpc, epithelial budding starts branching morphogenesis. Multipotent progenitor cells, which can become endocrine cells, exocrine cells, or ductal cells, exist at the tips of the branches. These cells are marked by *Pdx1*, *Ptf1a*, *carboxypeptidase A1* (*Cpa1*), and *myelocytomatosis oncogene* (*Myc* also known as *c-myc*).

Many genes have been shown to be involved in endocrine cell fate determination through gene knockout or transgenic mouse studies (for review, see 15). Genes implicated in specification of the four endocrine cell types are as follows: α cells, *POU domain class 3 transcription factor 4* (*Pou3f4*), *Foxa1*, *paired box gene 6* (*Pax6*), and *NK6 homeobox1* (*Nkx6.1*); β cells, *Pdx1*, *paired box gene 4* (*Pax4*), *NK2 transcription factor related locus 2* (*Nkx2.2*), *Nkx6.1*, *v-maf musculoaponeurotic fibrosarcoma oncogene family protein A* (*Mafa*), and *Foxa2*; somatostatin cells, *Pdx1* and *Pax4*; pancreatic polypeptide cells, *Nkx2.2* [15, 27, 40, 41]. The cell lineages and molecules involved in pancreatic development are summarized in Figure 2.

2.4 Intestine

2.4.1 Progenitors

The intestine endoderm, as specified from the endoderm and basic structure of the intestine, is established through epithelium-mesenchymal interaction [14]. Epithelium evaginates to form villi and intervillus regions. Undifferentiated and actively dividing cells exist in the intervillus region and invaginate to generate crypts.

2.4.2 Wnt signaling

The molecular mechanism of intestinal development is poorly understood compared to the intestinal adult stem cell. It is known that Wnt signaling has a key role in the intestinal development. Defects in cell proliferation and stability of the endodermal stem/ progenitor cells are observed in mice disrupted with the *transcription factor 7-like 2, T-cell specific,* and *HMG-box* (*Tcf7l2* also known as *Tcf4*) [18]. In this Wnt signaling–inhibited mutant mouse line, proliferative cells in the intervillus region are replaced by mature enterocytes.

At around E14.5, the earliest villus epithelial cells and crypt precursors are distinguished from each other by their marker genes. It is not until about E16.5 that the canonical Wnt signaling begins to work in the developing intestine, which is shown by the TOP-GAL reporter mice [17]. Early Wnt reporter activity is observed in the maturing epithelial cells lining the earliest villi, but not in the intervillus region [17]. In contrast, the crypt is well formed after birth by invasion of the intervillus region to the submucosa. After birth, canonical Wnt signaling works in the intervillus region. Therefore, Wnt signaling has a significant role in establishing the crypts, but not in the morphogenesis of the villus [38].

2.4.3 Hedgehog and BMP signalings

Ihh is expressed in the intervillus region, and *patched* (*Ptc*, hedgehog signaling receptor) is expressed in the adjacent mesoderm [21]. BMPs are expressed in the mesenchymal cells, and their expressions are regulated by the hedgehog signalings [3]. Ectopic crypt formation is generated by inhibiting BMP signaling, through

237

over-expression of noggin or conditional inactivation of *bone morphogenetic protein receptor, type 1A (Bmpr1a)*. These results suggest that BMP have a key role in restricting the number of the crypt [2, 12, 13].

2.4.4 Caudal type homebox 2 (Cdx2)

Recently it was reported that conditional *Cdx2*-deficient mice display a replacement of the posterior intestinal epithelium by the anterior esophagus epithelium without a major disruption in the expression of the *Hox* genes [9]. This result indicates that *Cdx2* regulates the identity and development of the intestinal progenitors, and this conversion is largely independent of the Hox code.

3. Adult Endodermal Stem Cells

Stem cells have unique characteristics: the ability to maintain themselves throughout long periods of time (self-renewal) and the capacity to generate all different cell types (multipotency) [24]. The capacity for self-renewal is assessed using bromodeoxyuridine (BrdU), which is incorporated into newly synthesized DNA. To clarify the multipotency of stem cells, *in vitro* clonal analysis or genetic lineage tracing experiments have been performed [24]. Late progenitor cells, which are known as transient amplifying cells [24], have restricted proliferation and differentiation potentials and are short-lived. In contrast, stem cells last longer and give rise to undifferentiated, intermediate, and differentiated progeny.

The main role of an adult stem cell is the maintenance and repair of its own tissue type. Adult stem cells have been defined in high cell turnover organs, including the skin, and low turnover organs such as the kidney and brain [8, 26,30]. Among the endodermally derived tissues, the liver and pancreas have a low cell turnover rate. The presence of liver adult stem cells and pancreatic adult stem cells has been discussed in several publications. On the other hand, the intestines exhibit a high turnover rate. Here we will focus on adult stem cells of the liver, pancreas, and intestine.

3.1 Adult liver stem cells

The adult liver plays an essential role in homeostasis, metabolism, and detoxification throughout adult life. These functions are carried out by hepatocytes, which constitute 80% of the liver cells. The liver shows a high regenerative potential after injury, which differs from a low turnover in the normal state. Thus, the liver-to-body mass ratio is maintained constant. The existence of liver stem cells is not clearly defined. Rapid regeneration of liver tissue is associated with replication of mature hepatocytes, and oval cells and/ or bone marrow cells are involved in regeneration under conditions when hepatocyte proliferation is suppressed [7]. Figure 3 shows the schematic drawing of hypothesized cell sources that are involved in liver regeneration.

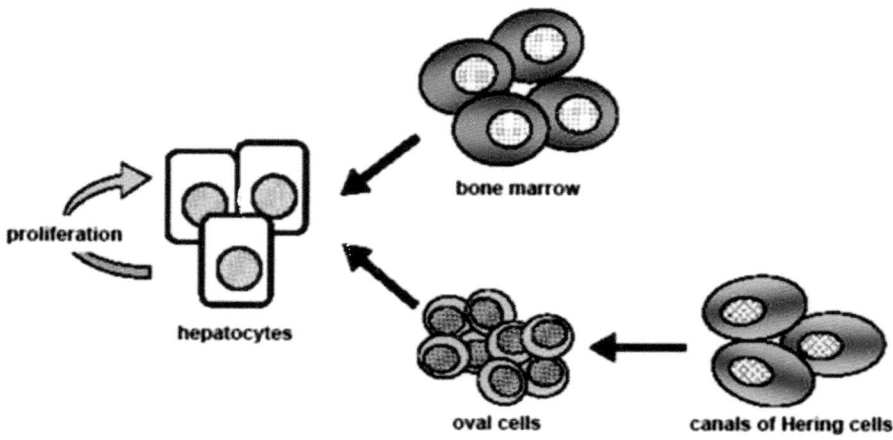

Figure 3. A schematic drawing of hypothesized cell sources that are involved in liver regeneration.

3.1.1 Replication of hepatocytes

Partial hepatectomy (PH) in the adult rat has been used to study hepatic regeneration. After 70% PH, normal liver tissue size recovered completely. Likewise, the human liver regenerates after PH and reaches its original size by 3–6 months. DNA synthesis in hepatocyte peaks at 24 hours after PH, and cell division is terminated by 2–3 days. Treatment with carbon tetrachloride (CCl_4) induces necrosis of hepatocytes and does not reduce liver mass. After CCl_4 treatment, the surviving hepatocytes replicate rapidly and maintain normal liver functions. These results indicate that hepatocytes can proliferate and contribute to regeneration of the liver.

3.1.2 Oval cells

Several methods are available to induce oval cells. Effective methods involve inhibiting proliferation of hepatocyte by exposure to 2-acetylaminofluorene (2-AAF), then subjecting the animals to PH to induce oval cell proliferation. Other approaches that do not include PH, such as treatment with a choline-deficient, ethionine-supplemented diet, induces oval cells in mouse liver. Oval cells have an ovoid nucleus and a high nucleus-to-cytoplasmic ratio. Oval cells express cholangiocyte markers of *keratin 19* (*Krt19*, also known as *cytokeratin 19*) and *gamma-glutamyltransferase 1* (*Ggt1*), a hepatoblast/immature hepatocyte marker of *Afp* and a hepatocyte marker of *Alb*. They also express hematopoietic stem cell markers such as *kit oncogene* (*Kit*, also known as *c-KIT*), *CD34 antigen* (*Cd34*), and *thymus cell antigen 1, theta* (*Thy1*). Oval cells are thought to give rise to ductal epithelial cells in the canals of Hering because these cells express both *Krt19* and *Afp* in the adult liver.

3.1.3 Bone marrow

Peterson et al. demonstrated that donor-derived bone marrow could repopulate the liver in rats treated with 2-AAF. Wang et al. reported that bone-marrow-derived hepatocytes could repopulate the liver of mice with fumarylacetoacetate hydrolase deficiency, which represents a model of tyrosinemia type 1 with marked liver disease. Furthermore, karyotyping demonstrated that the repopulated hepatocytes arose from cell fusion between host hepatocytes and donor-derived bone marrow cells.

3.2 Adult pancreatic stem cells

In the normal pancreas, few β cells replicate during adult life, and it is a slow cell turnover organ. The β cell mass is maintained during postnatal life in response to physiological damage and stresses such as disease, aging, and pregnancy. Physiological stimuli can induce the proliferation of β cells by 10-fold compared with the normal pancreas in rats. Likewise, a 30-fold increase in β cells was observed in mice resistant to insulin. Pancreatic regeneration models have shown that pancreatic duct ligation induces robust β cell hyperplasia. Moreover, pancreatitis induces exocrine damage; pancreatectomy induces regeneration of all pancreatic cell types; and streptozotocin (STZ) treatment induces regeneration of β cells. Here, previous reports, which suggested that β cells could be regenerated from progenitors within the islet, ducts, or acini, or by replication of the β cells themselves, are summarized. Figure 4 shows a schematic drawing of the cell sources for pancreas regeneration [5].

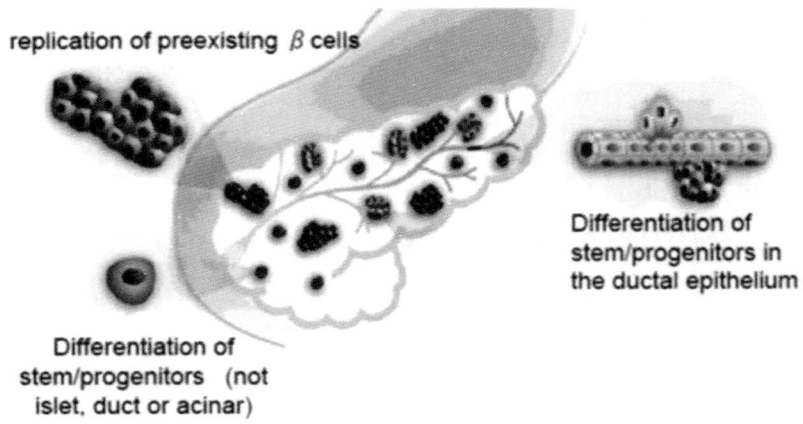

replication of preexisting β cells

Differentiation of stem/progenitors in the ductal epithelium

Differentiation of stem/progenitors (not islet, duct or acinar)

Adapted from Bonner-Weir S *et al*, 2005.

Figure 4. A schematic drawing of the cell sources for pancreas regeneration.

3.2.1 Progenitor cells within islets

Xu et al. demonstrated that the number of proliferating pancreatic cells increased 10-fold in a duct ligation model. The proliferating cells expressed *Neurog3*, which gives rise to all endocrine cell types, including β cells, thereby suggesting the existence of stem cells within islets. However, *Neurog3*-positive stem cells have not been identified in the normal adult pancreas. On the other hand, Thyssen et al. observed the appearance of proliferating cells expressing insulin or glucagon in the ducts, after a loss of β cells following STZ-treatment to neonatal rats [36]. Some glucagon-positive cells expressed *Pdx1*, thus suggesting they are of an immature α cell phenotype.

3.2.2 Progenitor cells in ducts

Bonner-Weir et al. reported that, following 90% pancreatectomy in the rat, a replication of preexisting endocrine or exocrine cells and an increase in the number of ductal structures occurred in focal regions. In the regenerating area, all duct cells expressed *Pdx1* transiently. Glucagon-positive cells and insulin-positive cells were observed, thereby suggesting that ductal cells might de-differentiate and transdifferentiate into endocrine cells. On the other hand, *carbonic anhydrase 2* (*Car2*, also known as *CAII*) is expressed in the pancreatic duct in the adult, but not in the embryonic tubular epithelium. Using a mouse line bearing a *Car2* promoter driving the expression of Cre recombinase (Car2-CreER™), in a duct ligation regeneration model, both endocrine and exocrine cells were observed as the progenies of *Car2*-positive mature duct cells.

3.2.3 β cell replication

It has been shown that βcells are themselves the main source of new β cells, throughout adulthood as well as during regeneration after pancreatectomy [35]. Mature β cells retain proliferative capacity, and expansion of adult β cell mass occurs by simple replication [6]. Using a transgenic mouse system for the specific and conditional ablation of β cells, Nir et al. reported that the surviving β cells were induced to proliferate during regeneration [25].

3.3 Small intestinal adult stem cell

In the intestine epithelium, differentiated cell types are classified by their functions: entercytes, goblet cells, enteroendocrine cells, and Paneth cells. The Paneth cells are located at the bottom of crypts, but major cells in the crypts are the so-called transit-amplifying cells. Intestinal epithelium renews every five days in the mouse small intestine. In this process, the transit-amplifying cells divide and rapidly differentiate into enterocytes, goblet cells, enteroendocrine cells, and Paneth cells. It is known that intestinal stem cells are located in the crypts. Active proliferation occurs within crypt compartment. The intestinal stem cells in the crypts proliferate in a Wnt-signaling dependent manner and can differentiate into enterocytes, goblet cells, and enteroendocrine cells [11]. This session focuses on the intestinal stem cell models and Wnt/β-catenin signaling pathway.

3.3.1 Small intestinal adult stem cells and the control by Wnt/β-catenin signaling pathway

Two stem cell models regarding the identity of the intestinal stem cells have been demonstrated: +4 *position* model and *stem cell zone* model [1] (Figure 5). In the +4 position model, stem cells are hypothesized to reside at the position +4 relative to the crypt bottom, with the first three positions occupied by terminal differentiated Paneth cells [32]. Potter et al. showed the existence of label-retaining cells residing specifically at +4 position. Recently, *Bmi1*, which play an essential self-renewal role in hematopoietic and neural stem cells, was reported to specifically mark undifferentiated cells at the bottom of crypts, located predominantly at the +4 position of crypts [32]. These cells expand and give rise into enterocytes, goblet cells, and enteroendocrine cells. In a ROSA26 conditional *Bmi1*-deficient mice, the intestines were devoid of crypts [32]. These results suggest that the *Bmi*-positive cells in +4 position represent small intestinal stem cells. However, it remains to be addressed if *Bmi1*-expressing cells overlap with a subpopulation of the intestinal stem cells described below.

In the stem cell zone model, crypt base columnar (CBC) cells represent the intestinal stem cells. Chen et al. demonstrated that CBC cells express *leucine-rich-repeat-containing G-protein-coupled receptor 5 (Lgr5)*, which is one of the Wnt targeted genes. *Lgr5*-deficiency induces premature differentiation of the Paneth cells. CBC cells can proliferate and differentiate into the enterocytes, golet cells, and enteroendocrine cells. Berker et al. indicated that CBC cells are the small intestinal stem cell by lineage tracing experiments. Sato et al. demonstrated that single Lgr5+ cells expanded, built crypt-villus epithelial structure, and differentiated into the goblet cells, Paneth cells, enterorndocrines and enterocytes [33]. These data strongly indicate that Lgr5+ cells represent intestinal stem cells.

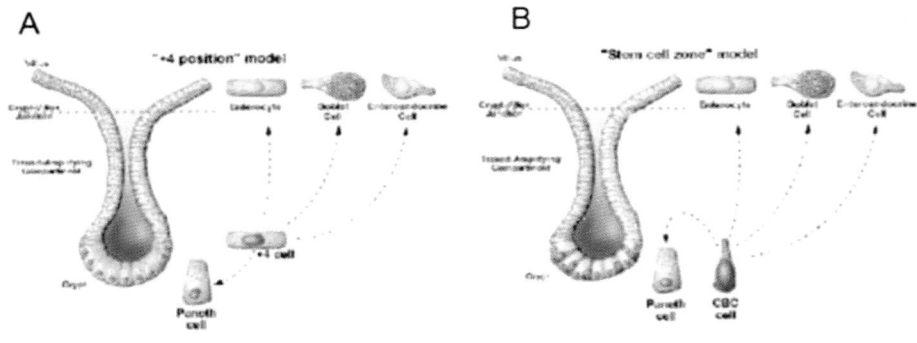

Figure 5. Schematic drawings of the intestinal stem cell models. +4 position model and stem cell zone model (modified from 1). +4 cell express *Bmi*. CBC cell express *Lrg5*.

In the absence of Wnt signaling, β-catenin are kept at a low level through a proteasomal degradation complex that consists of *adenomatous coli dedicate (APC), casein kinase I (CK I), glycogen synthase kinase 3 (GSK3),* and *axin*. Upon Wnt binding to the ligands of Frizzled on the cell surface, the inactivation of proteasomal degradation complex, Wnt was secreted by the epithelium at the crypts bottom. The activation of Wnt signaling induced the accumulation of nuclear β-catenin in the intestinal stem cells at the crypts. It was indicated that a gradient of Wnt signaling along the crypts-villus axis induced proliferation of the intestinal stem cells in the crypts. The transgenic mice, which ectopically express Wnt inhibitor *dickkopf homolog (Dkk)* regulated by the villin promoter, displayed a loss of proliferation. Taken together, Wnt/β-catenin signaling plays an important role in the maintenance and the proliferation of the crypts cells, including the intestinal stem cells.

The stem cell niche is required for the maintenance of self-renewal and multipotency. In the intestinal epithelium, intestinal stem cell niche is assumed to be provided by the stromal cells adjacent to the base of crypts. Yen et al. demonstrated that Wnt is expressed in the intestinal subepithelial stromal cells and Frizzled expressed in both the stromal cells and crypt epithelium. Ootani et al. reported that inhibition of Wnt caused the *Lgr5* positive cell growth and that the intestinal stem cell proliferation was dependent on stem cell niche using three-dimensional culture with Wnt agonist reproducing the niche [28]. These results indicated that the stromal cells promote the epithelial-mesenchymal crosstalk, which is required for maintaining the intestinal stem cell niche.

4. Conclusions

Here, we have reviewed the developmental regulation of embryonic precursor cells and recent understanding of the adult stem cells of endodermal tissues. There has been increasing interest in the therapeutic application of stem cells, albeit numerous obstacles remain. Knowledge from studies of specific gene-deficient mice and transgenic approaches has led to better understanding of the molecular mechanisms in embryonic development. Identification of tissue-specific adult stem cells and revealing the mechanisms of how their differentiation and self-renewal are controlled will provide breakthroughs for the successful manipulation of stem cells. In the next few decades, our knowledge will enable us to manipulate tissue stem cells in our bodies and apply them in clinical settings.

References

1. Barker, N., van de Wetering, M., and Clevers, H. 2008. The intestinal stem cell. *Genes Dev* 22:1856–64.

2. Batts, L. E., Polk, D. B., Dubois, R. N., and Kulessa, H. 2006. Bmp signaling is required for intestinal growth and morphogenesis. *Dev Dyn* 235:1563–70.

3. Bitgood, M. J. and McMahon, A. P. 1995. Hedgehog and Bmp genes are coexpressed at many diverse sites of cell-cell interaction in the mouse embryo. *Dev Biol* 172:126–38.

4. Bonal, C. and Herrera, P. L. 2008. Genes controlling pancreas ontogeny. *Int J Dev Biol* 52:823–35.

5. Bonner-Weir, S. and Weir, G. C. 2005. New sources of pancreatic beta-cells. *Nat Biotechnol* 23:857–61.

6. Dor, Y., Brown, J., Martinez, O. I., and Melton, D. A. 2004. Adult pancreatic beta-cells are formed by self-duplication rather than stem-cell differentiation. *Nature* 429, 41–6.

7. Fausto, N. and Campbell, J. S. 2003. The role of hepatocytes and oval cells in liver regeneration and repopulation. *Mech Dev* 120:117–30.

8. Galli, R., Gritti, A., Bonfanti, L., and Vescovi, A. L. 2003. Neural stem cells: an overview. *Circ Res* 92:598–608.

9. Gao, N., White, P., and Kaestner, K. H. 2009. Establishment of intestinal identity and epithelial-mesenchymal signaling by Cdx2. *Dev Cell* 16:588–99.

10. Grapin-Botton, A., Majithia, A. R., and Melton, D. A. 2001. Key events of pancreas formation are triggered in gut en-

doderm by ectopic expression of pancreatic regulatory genes. *Genes Dev* 15:444–54.

11. Haegebarth, A. and Clevers, H. 2009. Wnt signaling, lgr5, and stem cells in the intestine and skin. *Am J Pathol* 174:715–21.

12. Haramis, A. P., Begthel, H., van den Born, M., van Es, J., Jonkheer, S., Offerhaus, G. J., and Clevers, H. 2004. De novo crypt formation and juvenile polyposis on BMP inhibition in mouse intestine. *Science* 303:1684–6.

13. He, X. C., Zhang, J., Tong, W. G., Tawfik, O., Ross, J., Scoville, D. H., Tian, Q., Zeng, X., He, X., Wiedemann, L. M., et al. 2004. BMP signaling inhibits intestinal stem cell self-renewal through suppression of Wnt-beta-catenin signaling. *Nat Genet* 36:1117–21.

14. Hiramatsu, H. and Yasugi, S. 2004. Molecular analysis of the determination of developmental fate in the small intestinal epithelium in the chicken embryo. *Int J Dev Biol* 48:1141–8.

15. Jorgensen, M. C., Ahnfelt-Ronne, J., Hald, J., Madsen, O. D., Serup, P., and Hecksher-Sorensen, J. 2007. An illustrated review of early pancreas development in the mouse. *Endocr Rev* 28:685–705.

16. Katsumoto, K., Fukuda, K., Kimura, W., Shimamura, K., Yasugi, S., and Kume, S. 2009. Origin of pancreatic precursors in the chick embryo and the mechanism of endoderm regionalization. *Mech Dev* 126:539–51.

17. Kim, B. M., Mao, J., Taketo, M. M., and Shivdasani, R. A. 2007. Phases of canonical Wnt signaling during the development of mouse intestinal epithelium. *Gastroenterology* 133:529–38.

18. Korinek, V., Barker, N., Moerer, P., van Donselaar, E., Huls, G., Peters, P. J., and Clevers, H. 1998. Depletion of epithelial

stem-cell compartments in the small intestine of mice lacking Tcf-4. *Nat Genet* 19:379–83.

19. Lechler, T. and Fuchs, E. 2005. Asymmetric cell divisions promote stratification and differentiation of mammalian skin. *Nature* 437:275–80.

20. Lemaigre, F. P. 2009. Mechanisms of liver development: concepts for understanding liver disorders and design of novel therapies. *Gastroenterology* 137:62–79.

21. Madison, B. B., Braunstein, K., Kuizon, E., Portman, K., Qiao, X. T., and Gumucio, D. L. 2005. Epithelial hedgehog signals pattern the intestinal crypt-villus axis. *Development* 132:279–89.

22. Marshman, E., Booth, C., and Potten, C. S. 2002. The intestinal epithelial stem cell. *Bioessays* 24:91–8.

23. Matsuura, K., Katsumoto, K., Fukuda, K., Kume, K., and Kume, S. 2009. Conserved origin of the ventral pancreas in chicken. *Mech Dev*.

24. Morrison, S. J. and Spradling, A. C. 2008. Stem cells and niches: mechanisms that promote stem cell maintenance throughout life. *Cell* 132:598–611.

25. Nir, T., Melton, D. A., and Dor, Y. 2007. Recovery from diabetes in mice by beta cell regeneration. *J Clin Invest* 117:2553–61.

26. Oliver, J. A., Maarouf, O., Cheema, F. H., Martens, T. P., and Al-Awqati, Q. 2004. The renal papilla is a niche for adult kidney stem cells. *J Clin Invest* 114:795–804.

27. Oliver-Krasinski, J. M. and Stoffers, D. A. 2008. On the origin of the beta cell. *Genes Dev* 22:1998–2021.

28. Ootani, A., Li, X., Sangiorgi, E., Ho, Q. T., Ueno, H., Toda, S., Sugihara, H., Fujimoto, K., Weissman, I. L., Capecchi, M. R., et al. 2009. Sustained in vitro intestinal epithelial culture within a Wnt-dependent stem cell niche. *Nat Med* 15:701–6.

29. Potten, C. S. and Loeffler, M. 1990. Stem cells: attributes, cycles, spirals, pitfalls and uncertainties. Lessons for and from the crypt. *Development* 110:1001–20.

30. Potten, C. S. and Morris, R. J. 1988. Epithelial stem cells in vivo. *J Cell Sci Suppl* 10:45–62.

31. Rossi, J. M., Dunn, N. R., Hogan, B. L., and Zaret, K. S. 2001. Distinct mesodermal signals, including BMPs from the septum transversum mesenchyme, are required in combination for hepatogenesis from the endoderm. *Genes Dev* 15:1998–2009.

32. Sangiorgi, E. and Capecchi, M. R. 2008. Bmi1 is expressed in vivo in intestinal stem cells. *Nat Genet* 40:915–20.

33. Sato, T., Vries, R. G., Snippert, H. J., van de Wetering, M., Barker, N., Stange, D. E., van Es, J. H., Abo, A., Kujala, P., Peters, P. J., et al. 2009. Single Lgr5 stem cells build crypt-villus structures in vitro without a mesenchymal niche. *Nature* 459:262–5.

34. Sherley, J. L. 2002. Asymmetric cell kinetics genes: the key to expansion of adult stem cells in culture. *Stem Cells* 20:561–72.

35. Teta, M., Rankin, M. M., Long, S. Y., Stein, G. M., and Kushner, J. A. 2007. Growth and regeneration of adult beta cells does not involve specialized progenitors. *Dev Cell* 12:817–26.

36. Thyssen, S., Arany, E., and Hill, D. J. 2006. Ontogeny of regeneration of beta-cells in the neonatal rat after treatment with streptozotocin. *Endocrinology* 147:2346–56.

37. van den Brink, G. R. 2007. Hedgehog signaling in development and homeostasis of the gastrointestinal tract. *Physiol Rev* 87:1343–75.

38. Verzi, M. P. and Shivdasani, R. A. 2008. Wnt signaling in gut organogenesis. *Organogenesis* 4:87–91.

39. Weissman, I. L., Anderson, D. J., and Gage, F. 2001. Stem and progenitor cells: origins, phenotypes, lineage commitments, and transdifferentiations. *Annu Rev Cell Dev Biol* 17:387–403.

40. Zaret, K. S. 2008. Genetic programming of liver and pancreas progenitors: lessons for stem-cell differentiation. *Nat Rev Genet* 9:329–40.

41. Zaret, K. S. and Grompe, M. 2008. Generation and regeneration of cells of the liver and pancreas. *Science* 322:1490–4.

Chapter 10

iPS: Bioengineering Pluripotent Stem Cells

Timothy J. Nelson, Almudena Martinez-Fernandez, Satsuki Yamada, Andre Terzic

Division of Cardiovascular Diseases
Department of Medicine
Mayo Clinic
Rochester, MN

Nuclear reprogramming technology aims to reset the fate of ordinary cells back to a primitive stem cell ground state and generate advanced tools for unlimited regenerative medicine applications. By developing autologous sources of de novo stem cell pools, nuclear reprogramming offer a transformative approach that goes beyond the limits of natural stem cells to provide the indefinite potential of induced pluripotent stem (iPS) cells. In this way, patient-specific tissues have been obtained through bioengineered approaches independent of health or diseased conditions. Optimizing the derivation of iPS progeny will ensure a robust platform for novel discovery, diagnostics, and therapeutics not previously available due to the limitations of naturally derived stem cell platforms.

Here, the complete process of nuclear reprogramming of ordinary somatic tissues into bona fide pluripotent stem cells that were previously restricted to embryonic sources will be illustrated through state of the art technologies. The uniqueness of bioengineered stem cells resides in their autologous pluripotent capacity, which allows differentiation of all tissues and cell types that are genetically identical to the original starting source. The bioengineered cells have theoretically acquired an unlimited ability to replace virtually all tissues of the adult body, which significantly adds therapeutic value and establishes new paradigms for the practice of regenerative medicine. Therefore, the field of regenerative medicine is primed to adopt and incorporate iPS cell–based technology as a next generation stem cell platform to span the discovery, diagnostic, and therapeutic continuum across comprehensive medical and surgical specialties.

1. Stem Cell–Based Platforms

Stem cells are defined as progenitors derived from both natural and bioengineered processes that undergo asymmetric cell division in order to indefinitely maintain a clonal population while providing daughter cells capable of tissue-specific differentiation (table 1). The prototypical examples of naturally derived sources include embryonic, umbilical cord-derived, and adult stem cells. The bioengineered sources have produced stem cells through nuclear reprogramming that includes therapeutic cloning and ectopic transgenetic manipulation. Collectively, the various platforms of stem cells have unique advantages and distinct disadvantages. These variables typically focus on the source and availability of stem cell pools, the degree of plasticity or potential for tissue-specific differentiation, risk for malignant transformation, and immunologic status of stem cells compared to transplanted host environments.

Table 1. Stem cell–based platforms.

PLATFORM	DEFINING CHARACTERISTICS
Embryonic Stem Cell	• Derived from the inner cell mass of blastocyst • Pluripotent and self-renewing • Responsive to environmental cues
Perinatal Stem Cell	• Derived from perinatal sources • Mixture of embryonic-like and adult-like stem cells • Large supply available at time of birth
Adult Stem Cell	• Derived commonly from bone marrow, adipose tissues • Include resident stem cells • Multilineage clinical grade progenitors
Bioengineered Stem Cell	• Derived by reprogramming ordinary tissue sources • Include therapeutic cloning and nuclear reprogramming • Produce patient-specific embryonic-like stem cells

1.1 Natural stem cells

Naturally derived stem cells, including embryonic, umbilical cord blood, and adult stem cells, contribute to organ development in utero and tissue-renewal throughout adulthood [55,59,71,97]. Importantly, applications of natural stem cells have been utilized in the clinical setting for multiple applications and a wide spectrum of diseases, depending on the unique characteristics of the stem cells [15]. Such use of natural stem cells has provided a foundation for stem cell–based regenerative medicine [55–57]. Furthermore, this platform of natural stem cells has provided over the past few decades a fundamental foundation for stem cell–based regenerative medicine. The concept of pluirpotency and the innate limitation of natural stem cell populations provided a framework to define the differentiation capacity for all stem cell subpopulations. From the investigation of the molecular underpinning of natural pluripotency, candidate genes and pathways have been identified and used as agents of modulation through ectopic expression in non–stem cells to induce pluripotency and bioengineer a pluripotent state.

1.1.1 Embryonic stem cells

Embryonic stem (ES) cells are derived from the inner cell mass of pre-implantation blastocysts [78,89]. These original pluripotent stem cells give rise to all tissues of the adult body, including the complete repertoire of mesoderm, endoderm, and ectoderm derivatives during embryonic development. Over the past decade, a wide spectrum of ES cell lines have been produced across species, ranging from mouse to human. In particular, human ES cells were first isolated in 1998 and have demonstrated pluripotency with an innate ability to produce hematopoietic lineages, neurons, hepatocytes, pancreatic islets, osteocytes, chondrocytes, adipocytes, and cardiac progenitor cells. This quintessential differentiation capacity established a promising avenue for regenerative medicine with production of large quantity of transplantable cells from a theoretically renewable source.

Embryonic stem cells have been categorized as unique stem cells due to their unlimited self-renewal ability in cell culture, high nucleus to cytoplasmic ratio, and unrestricted differentiation that is capable of recapitulating the whole spectrum of normal embryonic development. The transcription factor *Oct4* coupled with *Sox2* and *E1A* contribute to the core transcriptional machinery responsible for maintenance of an undifferentiated ground state through fibroblast growth factor-4, WnT, and transforming growth factor-β dependent pathways [76]. Pluripotency ground state in mouse ES cells has been promoted with leukemia inhibitory factor through a STAT-3 dependent mechanism. Pluripotency in human ES cells is dependent on a Src-family of nonreceptor tyrosine kinases, and is validated by expression of alkaline phosphatase, POU transcription factor Oct3/4, Nanog, Cripto/TDGF1, proteoglycans TRA-1-60/81, GCTM-2, and embryonic antigens SSEA-3 and SSEA-4.

The major limitation of embryonic stem cells in regenerative applications relates to their risk of uncontrollable growth and teratoma formation that is inherent to the pluripotent ground state [78]. To remove this risk, progress in mapping and manipulating embryonic developmental pathways has established successful strategies to confine ES cells into tissue-specific lineages [58]. Ensuring the safety and advancement of new clinical applications requires proper regulation of ES cell differentiation before transplantation into the microenvironment of host tissue [5,52]. Thus, guidance of targeted differentiation using growth factors, cytokines, endogenous hormones, and small inorganic molecules provides a pharmacoinvasive approach to promote clinical-grade lineage specification. With the discovery of specialized biomarkers, undifferentiated progenitor cells can also be purged from differentiated progeny to allow additional security for lineage specification promoting targeted applications.

Independent of pluripotent growth potential, embryonic stem cells create a unique immunological challenge [35]. Since ES cells are derived from non-self embryonic tissues, engraftment into host tissue is by definition an allogeneic transplantation. Despite the inherent mismatch between host and transplanted stem cells, ES cells are capable of engraftment with minimal immunosuppression. Allogenic engraftment is thought to be due to low expression of MHC-1 protein in ES cells, and allows prolonged survival for allogeneic transplantations. Isotyped human leukocyte antigen (HLA) ES cell lines could thus be produced and characterized to generate a cell bank for "off-the-shelf" product development for individual patients in need of a stem cell source for regenerative therapies. Therapeutic applications of ES cells have so far been limited to pre-clinical studies with only one clinical trial approved by the U.S. Food and Drug Administration to date. The pre-clinical studies include neurological disorders, such as Parkinson's disease, and spinal injury [22] endocrine disorders, such as type-1 diabetes [36], cardiovascular disease, including ischemic and non-ischemic cardiomyopathy [20,37,98].

1.1.2 Perinatal stem cells

The differentiation capacity of perinatal stem cells, such as umbilical cord blood (UCB), has propelled this platform from discovery sciences into a clinically viable source for regenerative medicine. UCB is collected at the time of birth and provides a valuable pool of readily available stem cells that is commonly being stored in biobanks [91]. Transplantation of UCB has been clinically successful for hematopoietic stem cell applications resulting in a high degree of engraftment, favorable immunotolerance, and limited evidence for graft-versus-host disease compared to adult bone marrow stem cell transplantation.

Reconstitution of the adult hematopoietic system according to sibling donor transplant or by self-derivation was the initial rationale for UCB-based applications. In addition to hematopoietic stem cells, UCB-derived stem cells are capable of in vitro expansion, long-term maintenance, and differentiation into representative cells of all three embryonic germinal layers: endoderm (e.g., hepatopancreatic precursor cells, mature hepatocytes, type II alveolar pneumocytes), mesoderm (e.g., adipocytes, chondrocytes, osteoblasts, myocytes, endothelial cells), and ectoderm (e.g., neurons, astrocytes, oligodendrocytes) [91]. The heterogeneous populations of stem cells contained in UCB can be fractionated to purify a more homogeneous population with characteristics of embryonic-like stem cells. In this way, UCB provides a clinically applicable stem cell pool that avoids ethical challenges raised by embryonic sources [16].

Alternatively, amniotic epithelial cells (AEC) derived from amniotic membranes can also be induced to differentiate into diverse and specialized cell types from all three germ layers including pancreatic cells (endoderm), cardiomyocytes (mesoderm), and keratinocytes (ectoderm). Like UCB-derived cells, AEC are derived from pre-existing tissue that is readily available at the time of birth. Further advances with perinatal stem cells creates an opportunity to generate additional multi-lineage stem cells from non-embryonic sources.

1.1.3 Adult stem cells

Adult stem cells produce progenitors from a wide range of mature tissues such as bone marrow, adipose tissue, and resident stem cell pools [93]. Clinical applications of adult stem cells have lead the utilization of stem cells based on accessibility, autologous status, and favorable proliferative potential. Sufficient to recapitulate the entire hematopoietic system and provide mesenchymal stem cells with non-hematopoietic differentiation potential, bone marrow–derived stem cells are a cornerstone of contemporary adult stem cell applications [67].

Bone marrow–derived hematopoietic stem cells represent the earliest example of cell-based regenerative medicine, pioneered to alleviate the needs of patients treated with lethal doses of total-body irradiation for leukemia who developed life-threatening infections and irreversible tissue destruction. Stem cells defined by expression of the CD34 surface marker can be obtained via peripheral blood leukapheresis for clinical engraftment. Hematopoietic stem cells have provided the foundation for autologous and allogeneic stem cell transplantation, and offer novel treatments for patients with cancer, autoimmune diseases, and genetic diseases, including severe combined immunodeficiency and thalassemia. These transplant studies have importantly revealed engraftment of non-hematopoietic cell lineages derived from donor bone marrow, unmasking subpopulations capable of a diverse range of lineage-specific differentiation.

Mesenchymal stem cells were additionally discovered in the bone marrow, albeit at low frequency compared to the hematopoietic pool. Mesenchymal stem cells represent ~1 out of 10,000 nucleated bone marrow cells and arise from the supporting architecture of the adult marrow [13]. Comparable stem cells have also been isolated from connective components of various postnatal tissues including adipose and synovial tissue, as well as from peripheral and cord blood. Mesenchymal stem cells exhibit properties of multipotency (Table 2), with the capacity to contribute to regeneration of tissues of mesodermal origin, such as bone, cartilage, and muscle [70]. Evidence also supports the contribution of mesenchymal stem cells to liver and pancreatic islet cell regeneration, and protection in the setting of kidney, heart, or lung injury. In fact, both autologous and allogeneic mesenchymal stem cells have been tested in recent clinical trials including for treatment of osteogenesis imperfecta, Crohn's disease, and graft-versus-host disease [12].

Table 2. Stem cell differentiation potential.

POTENTIAL	DEVELOPMENTAL CAPACITY
Pluripotent	Ability to form all lineages of the body
Multipotent	Ability to form multiple cell types within a specific lineage
Unipotent	Ability to form a single cell type

Adult human mesenchymal stem cells are characteristically devoid of human leukocyte antigen (HLA) and class II antigens (MHC-II) on the cell surface, and do not express the co-stimulatory molecules CD80, CD86, or CD40 [13]. The expressed major histocompatibility complex (MHC) class I antigens may activate T cells, but with the absence of co-stimulatory molecules, a secondary signal would not engage, leaving T cells anergic. Such a unique immune profile (i.e., MHC I$^+$, MHC II$^-$, CD40$^-$, CD80$^-$, CD86$^-$), is regarded as non-immunogenic, and accordingly transplantation into an allogeneic host may not require immunosuppression [13]. Moreover, mesenchymal stem cells exhibit immunosuppressive properties modulating T cell functions, including cell activation, and display immunomodulatory features impairing maturation and function of dendritic cells and inhibiting human B-cell proliferation, differentiation, and chemotaxis. However, rejection of marrow stromal cells in MHC class I- and class II-mismatched recipients has been reported, underscoring the relevance of the immune response in the outcome of stem cell therapy [38].

At present, the use of allogeneic mesenchymal stem cell therapy in the management of acute disorders, such as acute myocardial infarction, acute graft-versus-host disease, and acute exacerbations of inflammatory bowel disorders has been considered. This strategy avoids the need for preparing autologous cells from the recipient. For disorders where mesenchymal stem cells are not needed on an emergency basis, it may be preferable to expand in vitro autologous mesenchymal stem cells prior to implantation in an effort to personalize cell-based therapy [38]. Expansion offers the opportunity to produce a large pool of naive stem cells, and derive progeny honed for lineage-specification away from multipotency and into tissue-restricted cytopoiesis. Derivation and characterization of specialized mesenchymal stem cell subpopulations, and their selective application to specific disease conditions is an emerging strategy for enhanced therapeutic outcome.

Collectively, natural stem cell platforms have been essential to uncovering fundamental mechanisms of self-maintanence and asymmetrical cell division that are required for continous production of differentiating progeny. Building on this foundation of knowledge, recent advances have enabled an entirely new platform for stem cells to rapidly emerge with unique characteristics not previously defined in natural stem cells.

1.2 Bioengineered stem cells

Exploiting epigenetics to influence the phenotypic outcome, biotechnology platforms reverse the differentiation state of ordinary cells, such as dermal fibroblasts, to achieve reprogramming of the cellular blueprint similar to that of an embryonic stem cell. Such platforms include therapeutic cloning and nuclear reprogramming that bypass the need for embryo extraction to generate pluripotent stem cell phenotypes from autologous sources (table 1) [28]. Reprogramming adult stem cells to generate customized embryonic-like stem cells offers the future for patient-specific regenerative therapies.

1.2.1 Therapeutic cloning through somatic cell nuclear transfer

Somatic cell nuclear transfer (SCNT) allows trans-acting factors present in the mammalian oocyte to reprogram somatic cell nuclei to an undifferentiated state (figure 1). Therapeutic cloning refers to SCNT in which the nuclear content of a somatic cell from an individual is transferred into an enucleated donor egg to derive blastocysts that contain pluripotent embryonic-like stem cells. From this approach, SCNT has produced cloned embryonic stem cells from multiple mammalian somatic cell biopsies [21,100]. The pluripotency of derived cells has been confirmed through germline transmission, and reproductive cloning. The robustness of this technology was originally highlighted by the successful production of Dolly the sheep in 1996 [11]. This scientific breakthrough was the first demonstration that the somatic cell nucleus retains the capacity to undergo reprogramming in the embryonic environment and is able to execute the proper developmental pathways characteristic of embryonic stem cells. Thus, in theory, it became possible to envision a bioengineering strategy to produce patient-specific stem cells for the purpose of therapeutic applications.

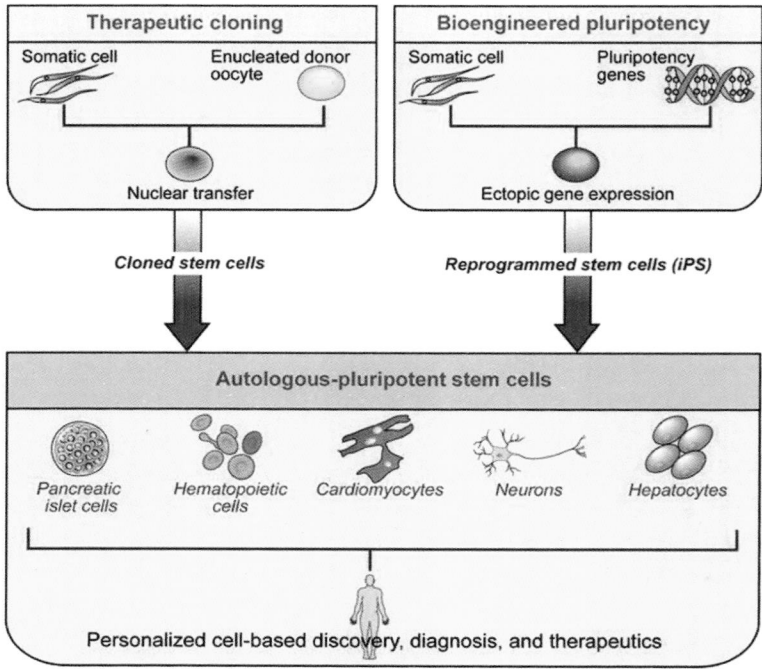

Figure 1. Bioengineered pluripotent stem cells. Pluripotent stem cells that function like embryonic stem cells can be produced from ordinary cells through therapeutic cloning or nuclear reprogramming. Therapeutic cloning combines the nuclear content of a somatic type with the cytoplasm of an enucleated donor oocyte. Transfer of the somatic cell nucleus into the remnant of the fertilized egg is performed by micromanipulation. This process results in the development of a blastocyst to allow harvest of pluripotent stem cells from the inner cell mass (ICM). Alternatively, similar starting somatic cell source can undergo nuclear reprogramming through ectopic expression of four genes (*Oct4* and Sox2 with either *Nanog* and *LIN28* or *Klf4* and *c-MYC*) to produce pluripotent cells independently of embryonic tissues. The bioengineered cells are called induced pluripotent stem (iPS) cells. Importantly, the progeny and tissues derived from engineered pluripotent stem cells are genetically similar to the original somatic cell biopsy and allow the advent of autologous, embryonic-like stem cells toward individualized cell-based applications.

1.2.2 Nuclear reprogramming

Bioengineering or nuclear reprogramming of adults cells (*sources*) exposed to pluripotency-associated transcription factors (*inductors*) in conjunction with complementary strategies to optimize the process (*facilitators*) is an attractive approach to induce an embryonic stem cell–like phenotype (figure 2). This strategy demonstrated the ability of a finite number of transgenic factors to recapitulate nuclear reprogramming in the absence of any embryonic tissues [85]. This breakthrough enabled the rapid adaption of nuclear reprogramming strategies by a wide spectrum of investigators and a diverse array of scientific questions geared not only to therapeutic applications but also fundamental discovery pipelines and individualized diagnostics.

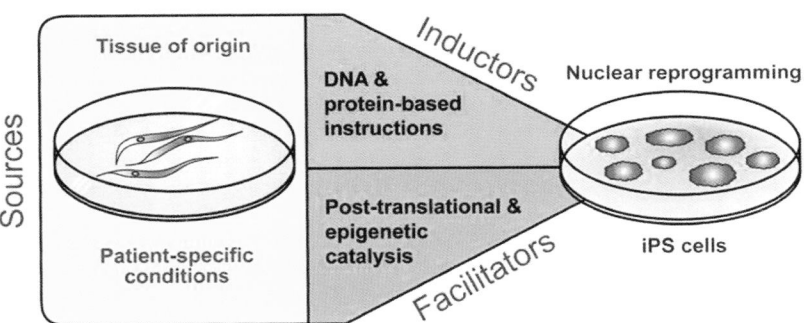

Figure 2. Toolkit for inducing pluripotency. Sources of parental cells provide the initial platform for nuclear reprogramming. Various tissues have been used for successful reprogramming in which the original identity of the starting cells may provide constructive influence for the process of reprogramming and subsequent differentiation. Inductors of nuclear reprogramming drive the reversible process of re-establishing a pluripotent ground state through downstream regulation of gene expression, proteome remodeling, conversion of metabolic machinery, and restructuring of the cyto-architecture. Facilitators provide a complementary axis to target the inherent barriers that limit the efficiency of successful nuclear reprogramming, and may be unique or universal to the parental cell types. Successful interactions between the three principal

components of the bioengineering toolkit allow derivation of functional pluripotency in generated progeny to be self-sustained in order to reset the fate of parental cells.

In the mouse, nuclear reprogramming approaches have reproducibly yielded induced pluripotent stem (iPS) cells sufficient for de novo embryogenesis and complete generation of germline stem cells [7,48,66,109]. In humans cells, the transcription factors sets, Oct4, Sox2, c-MYC, and Klf4 or alternatively Oct4, Sox2, Nanog, and Lin28, are sufficient to reprogram somatic tissue into pluripotent stem cells that exhibit the essential characteristics of ES cells. These include the maintenance of the developmental potential to differentiate engineered stem cells into advanced derivatives of all three germ layers [84,105]. Advantageously, iPS cell lines should largely eliminate the concern of immune rejection given the autologous immunological status of self-derived stem cells. Moreover, iPS-based technology will facilitate the production of cell line panels that closely reflect the genetic diversity of a population, enabling the discovery, development, and validation of therapies tailored for each individual.

2. Bioengineering Toolkit

The discovery that ectopic gene expression of a stemness-related gene set was able to induce cellular metamorphosis and convert ordinary cells into stem cell–like colonies was immediately recognized for its great potential for regenerative medicine and basic developmental biology. The resulting stem cell type was called an induced pluripotent stem (iPS) cell due to the ability to fulfill pluripotency criteria including differentiation into tissue types of the three germinal layers and integration into the developing embryo (Table 2). The iPS cells were hailed as a breakthrough since this bioengineering technique would allow the derivation of pluripotent cells from an autonomous source, thus bypassing the need for embryonic tissue in the process.

2.1 Source of parental cells

The use of a somatic cell population used for nuclear reprogramming has made iPS cell technology an unparalleled approach since the original observations that mouse embryonic fibroblasts were able to give rise to stem cell–like colonies [85]. Two mutually independent interpretations were invoked following the presumed observation of cellular conversion and fate reversal of somatic tissue [99]. One interpretation was that the starting cell source produced a heterogeneous population of cell types that contained a primitive quiescent stem cell contamination selected for and enriched through the process of nuclear reprogramming. In other words, reprogramming was only awakening the rare population of residual stem cells without fundamentally converting the cell fate. The alternative interpretation favored the concept of de novo reprogramming of mature non-stem cells, driving the epigenetics back into a pluripotent ground state that represented a distinctive reversal from the stable equilibrium maintained in the original cell population. These two hypothesis crystallized two models, the *elite* and the *stochastic* models of nuclear reprogramming [99].

The current evidence supports both models and indicates that mature and fully-differentiated cell types are capable of undergoing transformation back to a pluripotent ground state in a random pattern, supporting the stochastic paradigm. However, immature cell types, such as early stage lymphoid cell types, have an inherent advantage to undergo nuclear reprogramming compared to mature stages of the same lineage, which supports the elite model of reprogramming [90]. Therefore, less-mature cell types are primed for an increased likelihood of successful nuclear reprogramming. This differential reprogramming capacity can be explained in part by distinct levels of innate stemness-related transcription factors within different starting cell populations. These observations have led to the ability to eliminate the need for exogenous augmentation of select transgenic factors [1,34]. Case in point, neural stem cells with naturally high levels of *Sox2* have been reprogrammed with the single factor Oct3/4 [33]. Thus, cell type–specific augmentation of the stemness-factor levels is able to customize the strategy of nuclear reprogramming depending on the starting cell source.

2.1.1 Tissue origin of iPS cells

The origin of somatic cells used for the reprogramming process is important, apart from the issue of nuclear reprogramming efficiency. Depending on tissue source and age of the donor, teratogenic potential of reprogrammed progeny varies when injected into immunodeficient animals [51]. For instance, the iPS cells derived from adult tail tip fibroblasts and adult hepatocytes demonstrate the highest risk of dysregulated teratoma formation. In contrast, iPS cells derived from either embryonic fibroblasts or stomach epithelium demonstrate the lowest propensity to form tumors, which is similar to embryonic stem cells [51]. In terms of infectivity, however, stomach epithelium and hepatocytes offer a significant advantage and the ability to bioengineer ectopic expression with safer delivery strategies [2]. Other considerations to initiate the reprogramming process include the efficiency, safety, and availability of donor tissues from the most practical starting source.

To date, seven different mouse tissues have demonstrated the ability to be reprogrammed into iPS cells (table 3).

Although mouse embryonic fibroblasts were originally used, adult somatic tissues, including dermal skin [24, 53], liver/stomach biopsy [2], pancreatic beta-cells [80], neural stem cells [34, 75], and hematopoietic stem cells [19], have all demonstrated similar efficiencies using standardized methodologies for nuclear reprogramming. Further, successful reprogramming of human tissues has also demonstrated an array of starting sources that now include three other tissues beyond dermal skin [84, 105]: namely, keratinocytes [1], adipose tissue [82], and peripheral blood [42] (table 4). Similar criteria for initial functional pluripotency within the collection of these iPS clones have been met, according to teratoma formation. However, evidence suggests that iPS cells retain a memory of their parental source despite full nuclear reprogramming [44], which may thus provide the molecular basis for intrinsic differences between iPS clones generated from different parental somatic sources.

Table 3. Murine sources of ordinary tissue for nuclear reprogramming.

Murine Tissue Source	Induction strategy	Pluripotent criteria	Original references
Embryonic fibroblasts	Retrovirus-Oct4, Sox2, Klf4, c-MYC	Gene expression, teratoma formation, chimeric embryos	85
Tail-tip fibroblasts	Retrovirus-Oct4, Sox2, Klf4, c-MYC	Gene expression, teratoma formation, chimeric embryos	24, 53

Murine Tissue Source	Induction strategy	Pluripotent criteria	Original references
Hepatocytes and gastric epithelial	Retrovirus-Oct4, Sox2, Klf4, c-MYC	Gene expression, chimeric embryos	2
beta-pancreatic cells	tet-inducible lentiviruse-c-MYC, Klf4, Sox2, and Oct4	Gene expression, teratoma formation, chimeric embryos	80
Neural stem cell	Retrovirus-Oct4, Klf4 +/- Sox2, c-MYC and Retrovirus- Oct4, Sox2, Klf4, c-MycT58	Gene expression, teratoma formation, chimeric embryos	34,75
Hematopoietic stem cells	doxycycline-inducible lentiviruse-Oct4, Sox2, Klf4 and cMyc	Gene expression, chimeric embryos	19
Lymphocytes	doxycycline-inducible lentiviruse-Oct4, Sox2, Klf4 and c-MYC	Gene expression, teratoma formation, chimeric embryos	23

Table 4. Human sources of ordinary tissue for nuclear reprogramming.

Human Tissue Source	Induction strategy	Pluripotent criteria	Original references
Dermal skin fibroblasts	Retrovirus-Oct3/4, Sox2, Klf4, and c-MYC and Lentivirus-Oct4, Sox2, Nanog, and LIN28	Gene expression, teratoma formation	84, 104
Keratinocytes	Retrovirus-Oct3/4, Sox2, Klf4, and c-MYC	Gene expression, teratoma formation	1
Adipose tissue	Lentiovirus-Oct3/4, Sox2, Klf4, and c-MYC	Gene expression, teratoma formation	82
Peripheral blood (CD34⁺)	Retrovirus-Oct3/4, Sox2, Klf4, and c-MYC	Gene expression, teratoma formation	42

2.1.2 Patient-specific iPS

The bioengineered iPS platform permits the generation of autologous stem cells from individuals independent of disease conditions or genetic mutations. This feature, unique to autologous pluripotent stem cells generated by induced pluripotent technology, allows patient-specific stem cells to serve as a comparative platform for discovery science that is poised to revolutionize in vitro physiological and pharmacological studies. iPS cells have been generated from both genetic and non-inheritable diseases (table 5) for Parkinson's disease, adenosine deaminase-severe combined immunodeficiency, Gaucher's disease, Shwachman-Bodian-Diamond syndrome, Duchenne muscular dystrophy, Becker muscular dystrophy, Down syndrome, Huntington's disease, Lesch-Nyhan syndrome [69], myeloproliferative disorders [101], amyotrophic lateral sclerosis [17], Fanconi anemia [72], type 1 diabetes [43], spinal muscular atrophy [18], and familial dysautonomia [39]. Therefore, nuclear reprogramming has been successful in health and disease from multiple sources of adult tissues, providing a foundation to enable new paradigms of discovery science, novel diagnostics, and potential autologous cell-based therapeutics.

Table 5. Patient-specific iPS.

Disease or Syndrome	Induction strategy	Tissue source	Original references
Fanconi anemia	Retrovirus- Oct4, Sox2, Klf4, c-myc (mouse)	Dermal fibro-blasts	72
Amyotrophic lateral sclerosis (ALS)	Retrovirus- Oct4, Sox2, Klf4, c-MYC	Dermal fibro-blasts	17
Type 1 diabetes	Retrovirus- Oct4, Sox2, Klf4	Dermal fibro-blasts	43
Adenosine deaminase-severe com-bined immu-nodeficiency	Retrovirus- Oct4, Sox2, Klf4, c-MYC	Dermal fibro-blasts	69
Gaucher dis-ease type III	Retrovirus- Oct4, Sox2, Klf4, c-MYC	Dermal fibro-blasts	69

Disease or Syndrome	Induction strategy	Tissue source	Original references
Duchenne muscular dystrophy	Retrovirus- *Oct4, Sox2, Klf4, c-MYC*	Dermal fibro-blasts	69
Becker muscular dystrophy	Retrovirus- *Oct4, Sox2, Klf4, c-MYC*	Dermal fibro-blasts	69
Down syndrome	Retrovirus- *Oct4, Sox2, Klf4, c-MYC*	Dermal fibro-blasts	69
Parkinson disease	Retrovirus- *Oct4, Sox2, Klf4, c-MYC*	Dermal fibro-blasts	69
Swachman-Bodian-Diamond syndrome	Retrovirus- *Oct4, Sox2, Klf4, c-MYC*	Bone marrow mesen-chymal cells	69
Huntington disease	Retrovirus- *Oct4, Sox2, Klf4, c-MYC*	Dermal fibro-blasts	69
Lesch-Nyhan syndrome (carrier)	Retrovirus- *Oct4, Sox2, Klf4, c-MYC*	Dermal fibro-blasts	69
myeloprolifer-ative disorders (MPDs)	Retrovirus- *Oct4, Sox2, Klf4, c-myc* (mouse)	Perpher-al blood-CD34+ cells	101
Spinal muscu-lar atrophy	Lentiviral- *Oct4, Sox2, Nanog* and *LIN28*	Dermal fibro-blasts	18
Familial dysau-tonomia (FD)	Lentiviral- *Oct4, Sox2, Klf4* and *c-MYC*	Dermal fibro-blasts	39

2.2 Inductors of pluripotency

The original ectopic gene set sufficient to reprogram somatic cell types into pluripotent stem cells was revealed following a candidate-based screen strategy using a short list of twenty-four genes known to be associated with pluripotency [85]. These genes were identified according to their expression profiles within stem cells, thus labeled stemness-related genes. All genes were expressed and validated in embryonic fibroblasts through retroviral transduction to screen for cellular characteristics unique to pluripotent stem cells. The search for a combination of genes that were sufficient to reprogram a fibroblast into stem cells was the goal. The results demonstrated a minimum number of necessary genes, that is, a functional quartet of *Oct3/4*, *Sox2*, *Klf4*, and *c-MYC* that induced cellular metamorphosis and acquisition of stem cell characteristics (Figure 3) [85]. The *Oct3/4* and *Sox2* genes encode for transcription factors essential in the maintenance of pluripotency in early embryos. Complementary to the pluripotent genes, *Klf4* and *c-MYC* promote self-renewal in order to acquire an essential characteristic that is typical of stem cells. This seminal work was rapidly validated, adapted, and translated into protocols ultimately amenable for disease-specific applications.

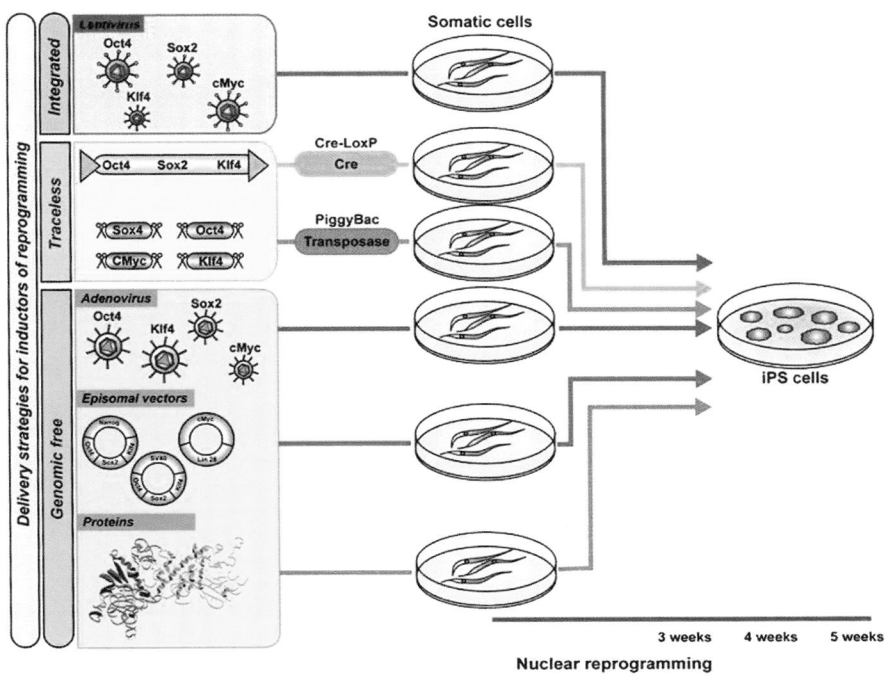

Figure 3. Candidate-based discovery approach for nuclear reprogramming. The original set of twenty-four stemness genes was distilled to four genes (*Oct4*, *Sox2*, *Klf4*, and *c-MYC*) that were sufficient to induce pluripotency from ordinary cell types. Alternatively, an independent screen of ninety-seven candidate genes revealed a second quartet of reprogramming factors that confirmed the robustness of *Oct4* and Sox2 in combination with *Nanog* and *LIN28*. Both sets of genes have been reproducibly applied and provide the gold standard for nuclear reprogramming platforms.

Independently, an alternative gene set sufficient for reprogramming functional pluripotency was identified according to similar candidate-based screening criteria that modified the quartet gene set. Both strategies prioritized *Oct3/4* and *Sox2* as core components, with *Klf4* and *c-MYC* being replaced with *Nanog* and *Lin28* in the alternative reported gene set [105]. The collection of pathways that these two sets of ectopic transgenes activate is likely similar, because *Lin28* and *c-MYC* demonstrate overlapping functions (figure 3). Both gene sets have been utilized in multiple parental cell types across a number of species to establish successful nuclear reprogramming strategies (tables 3 and 4). The robustness of iPS cell technology based on the potent inductors of reprogramming has reproducibly demonstrated the unique ability to generate stem cells that function like embryonic stem cells, even when cross-species genes and parental cell types are used [60]. The emerging trend is to match the specific stoichiometric needs of patient-derived cells with stemness-related factors to ensure the quality and efficiency of nuclear reprogramming.

2.3 Facilitators of reprogramming

The multifaceted influence of stemness-related gene sets induces a wide range of changes from gene expression networks to protein architecture remodeling and metabolic machinery retooling. Applying a complementary strategy to optimize the rate-limiting steps of nuclear reprogramming offers incremental advantages for the overall efficiency of deriving iPS cells from original parental cell sources.

Facilitators are small molecules or conditions designed to improve the efficacy of inductors to induce nuclear reprogramming. Among the facilitators, manipulation of the oxygen levels to 5% hypoxia compared to standard 21% oxygen creates a favorable environment that provokes a complex rearrangement of epigenetics to promote the induction of reprogramming [103]. Important to the epigenetic state, histone acetylation and methylation regulate the accessibility of the transcription machinery to the genetic blueprint within the parental cell. In this context, inhibition of histone deacetylase or

DNA methyltransferase using epigenetic modifiers, valproic acid or 5-aza-cytidine, has benefited chromatin remodeling and increased the efficiency of standardized reprogramming protocols [27,50].

Chemical modulation of signaling pathways involved in maintenance of the pluripotent state can significantly improve the overall efficiency of nuclear reprogramming as well [41]. The ubiquitous surveillance activity of *p53*, which functions as a tumor suppressor gene, has been identified as a critical rate-limiting roadblock during the early stages of nuclear reprogramming. Accordingly, temporary *p53* knockdown either by gene silencing or protein degradation increases the overall efficiency of successful progression through all stages of nuclear reprogramming [3,25,31,40,45,90]. Together, these strategies modify the susceptibility of the original somatic cells and boost mechanisms that regulate the checkpoints restricting conversion of cell fate through nuclear reprogramming.

3. Delivery Strategies for Nuclear Reprogramming

Nuclear reprogramming through triggered cellular dedifferentiation offers a revolutionary framework to derive pluripotent stem cells from somatic tissue, independent of an embryo source. The emerging technology demonstrates that ectopic expression of stemness-related genes is sufficient to reset parental cell fate, unlocking the potential for unlimited patient-specific regenerative therapies. Thereby, discovery science has overcome restrictions inherent to embryonic derivation enabling access to a genuine, autologous, pluripotent cell population. Nuclear reprogramming technology must, however, limit the footprint or residual modifications to the genome in order to establish a safe tool amenable for clinical translation (**Figure 4**).

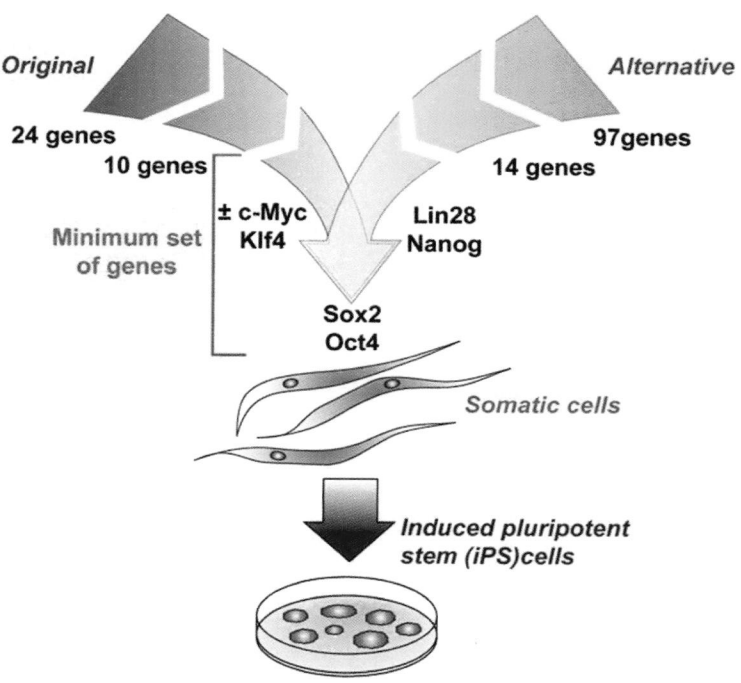

Figure 4. Strategies to deliver inductors of pluripotency. Genetic engineering with retrovirus and lentivirus provided the initial strategy to efficiently deliver adequate levels of ectopic transgenes to evaluate the process of nuclear reprogramming. The genomic modifications that are inevitable with use of integrating vectors create, however, a potential risk of insertional mutagenesis. Traceless engineering was therefore established to remove the risk of genomic modifications by enzymatically cutting the ectopic sequences out of the genome after successful nuclear reprogramming. Alternatively, genomic-modification free technology has established in proof of principle studies reprogramming with plasmids, episome vectors, and recombinant proteins.

3.1 Genetic engineering

Retroviral and lentiviral approaches offered the initial methodology that launched the field of gene delivery for nuclear reprogramming and established the technological basis of iPS with rapid confirmation from multiple vector systems. Because the retroviral-based vector systems have built-in sequences that silence the process of transcription upon pluripotent induction, ectopic gene expression was cut off as cells achieved a stable ground state [26]. This observation indicated that successful self-maintenance of the pluripotent state was possible independent of persistent transgene expression. Thus, transient expression of ectopic genes enabled the strategy of nuclear reprogramming to be reconsidered as a temporary exposure to stemness-related factors and promoted the notion of ectopic transgene removal once reprogramming was achieved. The random integration and permanent genetic alterations of exogenous DNA, inherent to retroviral or lentiviral constructions, was therefore not an essential component for next generation technology as the science moves toward safer clinical applications. For this reason, new approaches aim to identify the most potent stemness-related factors that are able to more efficiently interact with native cellular machinery and successfully rewrite the pluripotent pattern of gene expression without long-term disruption of the underlying genetic integrity of the host cell.

3.2 Traceless engineering

Two biphasic traceless systems have been employed to produce iPS cells independent of permanent genetic modifications. The classic Cre-LoxP method was engineered to delete ectopic sequences from the iPS genome upon successful nuclear reprogramming [14]. This so-called hit and run system removes residual sequences of the transgene but notably leaves a small LoxP site at the site of random integration.

The latest innovation that advances iPS-based technology towards clinical applications has been recently highlighted where non-viral approaches are capable of high-efficiency iPS production in truly traceless fashion [63]. These approaches are dependent on short sequences of mobile genetic elements that can be used to integrate transgenes into host cell genomes and provide a genetic tag to "cut and paste" flanked genomic DNA sequences. This piggyBac (PB) system couples enzymatic cleavage with sequence-specific recognition using a transposon/transposase interaction to ensure high efficiency removal of flanked DNA without residual genetic footprint [95]. Importantly, this technology achieves a traceless transgenic approach, where non-native genomic sequences that are transiently required for nuclear reprogramming can be removed upon induction of pluripotency. Using the PB transposition system with randomly integrated stemness-related transgenes, it has been demonstrated that disposal of ectopic genes can be efficiently regulated upon induction of self-maintaining pluripotency [95]. This state of the art system is uniquely qualified to allow safe integration and removal of ectopic transgenes, improving the efficiency of iPS production.

3.3 Genomic-modification free

The risk of oncogenic genes and insertional mutagenesis, inherent to stable genomic integration, triggered in turn the search for the next generation of iPS production. Thereby, systems were designed for transient production of stemness-related genes without integration into the genome. The first proof of principle was achieved by non-integrating viral vector systems, such as adenovirus [81], and confirmed by repeated exposure to extra chromosomal plasmid-based transgenes [65]. Importantly, these reports demonstrated that expression of stemness-related factors was required for only a limited time frame until progeny developed autonomous self-renewal, establishing nuclear reprogramming as a bioengineered process that resets a sustainable pluripotent cell fate independent of permanent genomic modifications. However, the inefficiency of non-integrated or genomic modification–free technologies has hindered broader applicability. Alternatively, the security of unmodified genomic intervention can be achieved with non-integrating episomal vectors [104]. Vectors are also being developed based on transient expression and methods to eliminate the risk of permanent insertional mutagenesis [29].

Finally, direct delivery of recombinant proteins engineered to penetrate across the plasma membrane of somatic cells and translocate into the nucleus has succeeded to be sufficient to induce pluripotency in somatic cells, initiating the new concept of DNA-independent reprogramming [109]. Two methodologies have been successful with recombinant protein technologies. First, by transiently permeabilizing the cell membrane of the parental fibroblasts, embryonic stem cell extracts have been successful in obtaining a reprogrammed stem cell population [8]. Second, bioengineering recombinant proteins with membrane-translocating peptide consisting of multiple arginine amino acids at the N-terminus of the stemness-related factors has enabled an alternative strategy [32, 109]. Collectively, these strategies for nuclear reprogramming accelerate the discovery science of regenerative medicine and bring the technology a step closer to clinical applicability. By producing genetically unmodified progenitor cells that acquire the capacity of pluripotency, advances in the field of nuclear reprogramming make it theoretically possible to generate unlimited, autologous tissues from patients for both diagnostic and therapeutic applications.

4. Stringency Tests of Pluripotent Competency

Originally, somatic cell nuclear transfer established the proof of principle concept of genomic intervention to reset cell fate. Transferring the nucleus of an adult cell into the cytoplasmic environment of a host enucleated oocyte revealed the malleability of mammalian somatic cells to acquire atavistic developmental competency [6,10,21,100]. Ensuing nucleus-to-cytosol interactions catalyze the epigenetic regulation that reverses the differentiation state of the parental nucleus to achieve genetic reprogramming and recapitulate the embryonic ground state. This provides tools to re-engineer the cellular make-up including the mitochondrial machinery of ordinary cells [83]. The robustness of this process is further evident by successful xenogenic combinations of adult nuclei with primitive cytoplasm components, revealing phylogenetic conservation of a permissive gene-environment interface [6]. Recognition that so-called terminally differentiated tissue phenotype is reversible to the pluripotent state paved the way towards embryo-independent engineering.

The discovery that induced pluripotency was achievable in response to ectopic expression of stemness-related transgenes required measurable milestones to establish criteria of embryonic trait re-acquisition, such as germline transmission that was previously only attained by natural embryonic stem cells. Bioengineered iPS clones need to meet multiple levels of pluripotent stringency to collectively validate the presumed primitive characteristics of embryonic stem cells, which is the gold standard. These criteria span from in vitro differentiation in cell culture to in vivo functional properties to in utero developmental potential and to ultimately in situ regeneration of diseased tissues (figure 5).

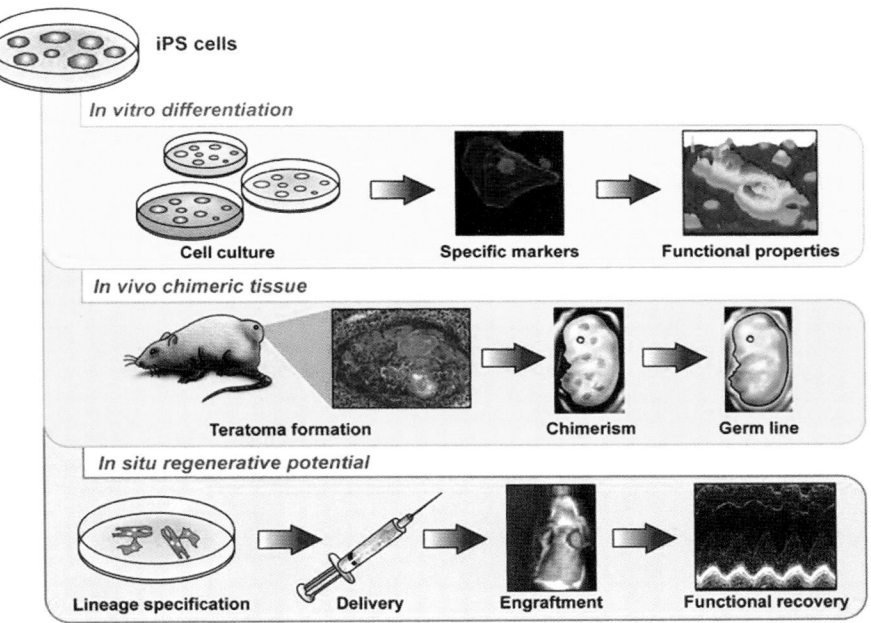

Figure 5. Stringency tests of pluirpotent competency. Nuclear reprogrammed fibroblasts must meet pluripotent stringency criteria in order to be classified as iPS cells. In vitro differentiation provides the initial characterization of bioengineered stem cells to demonstrate the acquisition of specialized cellular differentiation such as cardiomyocytes. In vivo chimeric tissue offer an alternative strategy to decipher reprogrammed progeny that have undergone cellular conversion and acquired pluripotent differentiation capacity either in immunodecificent teratoma assay or in utero embryonic development leading to germline transmission. In situ regeneration potential determines the ability of reprogrammed stem cells to respond to injured tissues or diseased environments and drive de novo cellular repair for restoration of healthy organ function.

4.1 In vitro differentiation

As potential pluripotent stem cells, iPS cells can be cultured in vitro and differentiated using existing protocols validated for embryonic stem cells. Differentiation of iPS cells into functional cell types is a necessary step toward their applicability in regenerative medicine and therapeutic screening. So far, embryonic stem cell–designed protocols have been used to differentiate iPS cells, giving rise to cardiomyocytes [46,47,54,73,108], adipocytes and osteoblasts [86], hematopoietic and differentiated blood cells [64], dendritic cells and macrophages [74], insulin-producing cells [43,87,107], hepatocyte-like cells [77, 79], retinal cells [9,49,68], and several types of neurons [17,18,30].

In all cases the identity of differentiated progeny can be confirmed using immunofluorescent staining for specific markers coupled with gene expression profiling. Furthermore, mature tissues derived from iPS cells, such as cardiomyocytes, neurons, hematopoietic lineages, or other specialized cell types, require an additional degree of specificity to characterize the functional properties of cell types needed to mimic the physiological behavior of natural tissue. For example, basal functionality of iPS-derived cardiomyocytes has been characterized with calcium-mediated electromechanical coupling, action potentials, and ion current components under physiological conditions [46,47]. From a more clinical standpoint, effects of cardioactive drugs on human iPS-derived cardiomyocytes have also been studied, revealing a similar response to that of embryonic stem cell–derived cardiac cells and matching the expected clinical effect for the tested molecules [102]. Also, motor neurons differentiated from reprogrammed cells have been studied, demonstrating that their excitability match characteristics expected from adult cells [30]. Furthermore, endoderm lineages of functional pancreatic beta-cells and liver parenchymal hepatocytes have also been recapitulated in vitro to provide the basis for further translation into diagnostic and therapeutic applications [43,77]. Thus, these data demonstrate the diversity of lineage specification and functionality that can be produced through de novo differentiation from bioengineered pluripotent stem cells.

4.2 In vivo chimeric tissue

Pluripotent cells have been defined by their innate ability to give rise to multi-lineage tumors when transplanted subcutaneously in immunodeficient hosts. Beyond artificial modulation of extracellular conditions, this in vivo model system provides a simple yet sensitive and efficient methodology to test the differentiation potential of bioengineered stem cells. Production of three germ layers in tumors, known as teratoma, ensures a histological high-throughput readout of the net outcome for multiple cell clones. Importantly, this analysis provides the most stringent criterion for human pluripotent or reprogrammed cells given the ethical limitations for alternative tests that can be applied to non-human stem cells.

Since embryonic stem cells are derived from the inner cell mass of a pre-implantation blastocyst, non-human pluripotent stem cells should be able to function in an equivalent fashion throughout chimeric embryonic development. As primordial cells, pluripotent stem cells are, in theory, able to contribute to all the tissues of the developing embryo. Reprogramming to iPS resets expression profile and characteristics to those of natural embryonic stem cells allowing their integration within pre-implantation embryos. To test this capacity, iPS cells of interest are incubated with eight-cell embryos (morula) or injected into a blastocyst and transferred back into the uterus of a stage-appropriate surrogate mother. If nuclear reprogramming has been successfully completed, iPS cells will integrate and populate the developing embryo to generate iPS-derived tissue in the chimeric offspring. In the event of healthy pluripotent cells, they will contribute not only to developing tissue, but also to the specialized germ cells that are capable of germline transmission to naturally breed offspring.

For non-human stem cells, the most stringent test for pluripotency is tetraploid complementation assay. In this assay, two-cell embryos are harvested to allow ex utero manipulation. First, the embryos receive a controlled electrical impulse sufficient to fuse the two cells into a single tetraploid cell that contains two copies (4N) of the genome. The genetically abnormal 4N tetraploid embryo subsequently cannot develop into the embryo proper, yet allows the growth of transient extraembryonic tissues required to support a developing fetus. However, if pluripotent stem cells that are functionally equivalent to natural embryonic stem cells are aggregated with the mutant 4N embryo, the chimeric embryo is able to execute normal developmental processes. The developing embryo and subsequent liveborn offspring are thus completely derived from the transplanted pluripotent stem cells, and thereby establish the highest-stringency for pluripotency in non-human stem cells [7,106].

4.3 In situ regeneration potential

iPS technology has overcome inherent restrictions to embryonic stem cells, enabling a bioengineered pluripotent stem cell derived from "self" to produce autologous tissues. In this way, the potential for ethical concerns with regard to embryonic tissues and immunological mismatch from non-autologous stem cells is essentially eliminated. Therefore, a unique advantage of self-repair of multiple tissues allows a new criterion for iPS cells, in the context of regenerative medicine.

Since multiple stem cell populations have demonstrated therapeutic repair following transplantation into diseased tissues, iPS cells have also been put to the task of in situ regeneration of disease models. Proof of principle applications for this strategy have been provided to-date in animal models addressing blood [24,96], neural [94], and cardiac disease [61] (table 6). The first treated model was a humanized sickle cell anemia mouse, from which iPS were derived, corrected for their genetic defect using homologous recombination and differentiated into healthy hematopoietic cells within a model system of severe disease phenotype [24]. When corrected iPS-derived hematopoietic cells were transplanted into the sickle cell anemia model, they engrafted properly and reversed the severe disease phenotype [24]. Similarly, deficit in coagulation characteristic of hemophilia A was shown to improve after injection of iPS-derived healthy endothelial cells into the liver of model animals [96]. With sufficient expression of factor VIII from iPS cells, several major organ systems were consequently repaired.

Table 6. Applications for iPS-based therapeutic repair.

Disease Condition	Therapeutic outcome
Sickle cell disease	Hematopoiesis, functional physiological improvement
Parkinson's disease	Dopamine production, symptomatic improvement
Hemophilia A	Decreased clotting time, survival benefit
Ischemic heart disease	Improved cardiac performance, *in situ* tissue repair

Moreover, in the case of Parkinson's disease, motor neurons obtained from iPS cells were demonstrated to engraft and integrate in the striatum of diseased rats after local injection. Functional engraftment significantly reduced the neurological symptoms in this model system [94]. Regarding cardiovascular disease, iPS injected directly into the myocardium in a mouse model of infarction showed sustained presence in the heart and significant improvement of cardiac performance [61]. Post-ischemic cardiac function was compared in randomized cohorts transplanted with parental fibroblasts versus bioengineered iPS cells. As quantified by echocardiography, occlusion of anterior epicardial coronary blood flow permanently impaired regional wall motion and cardiac function. Treatment with parental fibroblasts was unable to improve the structure and function of post-ischemic hearts throughout a four-week follow-up as demonstrated by coordinated concentric contractions visualized by 2-D imaging. In contrast, iPS treatment of ischemic heart failure improved functional performance, prevented structural remodeling, and avoided deleterious effects on rhythm propagation [61]. These examples set the experimental basis for an iPS-based curative approach to manage degenerative diseases that have previously been considered incurable by traditional management.

5. Applications for iPS-Based Technology

As medical therapy is moving from a palliative approach towards a curative algorithm for individual patients, regenerative medicine will become a priority to healthcare transformation [92]. In this regard, technological advances of induced pluripotent stem cells are driving a new scientific

288

platform that will be increasingly essential for realizing the opportunities of patient-specific, cell-based diagnostics and therapeutics [62].

5.1 Patient-specific diagnostics

The field of regenerative medicine is growing in conjuncture with the realization of the individualized medicine paradigm that aims to create predictive, personalized, and preemptive solutions for patient-specific diagnostics [88]. Individualized treatment algorithms for regenerative medicine are dependent on the initial qualification of the inherent reparative potential of the patients. In this way, the *stem cell load* for each patient will serve as an *index for regenerative potential* that will guide prediction, diagnosis, and prognosis of degenerative diseases that are due to a lack of natural healing processes [55,56].

Furthermore, patient-specific iPS cells provide a new platform for discovery science to reveal mechanisms of disease. By having an unlimited source of a patient's own cells in the form of a primordial stem cell, detailed mechanistics studies comparing health and disease offer a powerful strategy to dissect the molecular etiology of a disease independent of co-morbidities, drug therapy, age, and the heterogeneity of environmental factors. In the context of monogenetic diseases, multi-generational structured pedigrees have been able to map through linkage analysis disease-causing mutations. Patient-specific iPS cells provide a unique opportunity to directly study genotype/phenotype interactions. Well-controlled genetic populations with the benefit of detailed clinical information enable patient-specific iPS cells to contribute to the study of variable penetrance of presumed monogenetic disorders.

This novel combination of technologies may produce immediate benefits to the patient and their families through predictive diagnostics based on bioengineered pluripotent stem cells.

5.2 Therapeutic regeneration

The long-term goal of regenerative medicine that aims to repair and restore normal tissue structure and function in disease has been accelerated by the breakthrough in patient-specific iPS cells [57]. The obvious benefits of self-derived tissue eliminate the need of immunosuppression for cell-based transplantation. Cytotoxic medications that are required to inhibit the ability of the innate immune system to destroy transplanted allogeneic tissue have intense side-effects that long have been linked to the high occurrence of secondary malignancies and degenerative disease outcomes. Therefore, avoidance of immunosuppression offers a transformation for transplant medicine. Furthermore, creating bona fide pluripotent stem cells from patient-derived sources unlocks the restrictions of tissue-specific differentiation that limits non-pluripotent stem cells. Collectively, these two components of iPS cells expand the scope of therapeutic options. If applications of iPS cells in regenerative medicine are able to match the right person with the right progenitor cell at the right time, then the future of clinical practice may be able to transition away from the non-curative disease management models of today, and provide lasting treatments that reverse the degeneration of incurable diseases through personalized patient-specific solutions.

References

1.   Aasen T., Raya A., Barrero M. J., Garreta E., Consiglio A., Gonzalez F., Vassena R., Biliú J., Pekarik V., Tiscornia G., Edel M., Boué S., Izpisúa Belmonte J. C. 2008. Efficient and rapid generation of induced pluripotent stem cells from human keratinocytes. *Nat Biotechnol* 26:1276–1284.

2.   Aoi T., Yae K., Nakagawa M., Ichisaka T., Okita K., Takahashi K., Chiba T., Yamanaka S. 2008.Generation of pluripotent stem cells from adult mouse liver and stomach cells. *Science* 321:699–702.

3.   Banito A., Rashid S. T., Acosta J. C., Li S., Pereira C. F., Geti I., Pinho S., Silva J. C., Azuara V., Walsh M., Vallier L., Gil J. 2009. Senescence impairs successful reprogramming to pluripotent stem cells. *Genes Dev* 23:2134–2139.

4.   Behfar A., Faustino R. S., Arrell D. K., Dzeja PP, Perez-Terzic C., Terzic A. 2008. Guided stem cell cardiopoiesis: discovery and translation. *J Mol Cell Cardiol* 45:523–529.

5.   Behfar A., Perez-Terzic C., Faustino R. S., Arrell D. K., Hodgson D. M., Yamada S., Puceat M., Niederländer N., Alekseev AE, Zingman LV, Terzic A. 2007. Cardiopoietic programming of embryonic stem cells for tumor-free heart repair. *J. Exp Med* 204:405–420.

6.   Beyhan Z., lager AE, Cibelli JB. 2007. Interspecies nuclear transfer: Implications for embryonic stem cell biology. *Cell Stem Cell* 1:502–512.

7.   Boland M. J., Hazen J. L., Nazor K. L., Rodriguez AR, Gifford W., Martin G., Kupriyanov S., Baldwin KK. 2009. Adult mice generated from induced pluripotent stem cells. *Nature* 461:91–94.

8.   Bru T., Clarke C., McGrew M. J., Sang HM, Wilmut I., Blow J. J. 2008. Rapid induction of pluripotency genes after exposure of human somatic cells to mouse ES cell extracts. *Exp Cell Res* 314:2634–2642.

9.   Buchholz DE, Hikita S. T., Rowland T. J., Friedrich AM, Hinman CR, Johnson LV, Clegg DO. 2009. Derivation of functional retinal pigmented epithelium from induced pluripotent stem cells. *Stem Cells* 27:2427–2434.

10.   Byrne J. A., Pedersen DA, Clepper LL, Nelson M., Sanger WG, Gokhale S., Wolf D. P., Mitalipov SM. 2007. Producing primate embryonic stem cells by somatic cell nuclear transfer. *Nature* 450:497–502.

11.   Campbell KH, McWhir J., Ritchie WA, Wilmut I. 1996. Sheep cloned by nuclear transfer from a cultured cell line. *Nature* 380:64–66.

12.   Caplan AL. 2007. Adult mesenchymal stem cells for tissue engineering versus regenerative medicine. *J Cell Physiol* 213:341–347.

13.   Chamberlain G., Fox J., Ashton B., Middleton J. 2007. Mesenchymal stem cells: their phenotype, differentiation capacity, immunological features, and potential for homing. *Stem Cells* 25:2739–2749.

14.   Chang CW, Lai YS, Pawlik KM, Liu K., Sun CW, Li C., Schoeb TR, Townes TM. 2009. Polycistronic lentiviral vector for "hit and run" reprogramming of adult skin fibroblasts to induced pluripotent stem cells. *Stem Cells* 27:1042–1049.

15.   Copelan, E.A. 2006. Hematopoietic stem-cell transplantation. *N Engl J Med* 354:1813–1826.

16. De Coppi P., Bartsch G. Jr, Siddiqui MM, Xu T., Santos CC, Perin L., Mostoslavsky G., Serre AC, Snyder EY, Yoo J. J., Furth ME, Soker S., Atala A. 2007. Isolation of amniotic stem cell lines with potential for therapy. *Nat Biotechnol* 25:100–106.

17. Dimos JT, Rodolfa KT, Niakan KK, Weisenthal LM, Mitsumoto H., Chung W., Croft G. F., Saphier G., Leibel R., Goland R., Wichterle H., Henderson CE, Eggan K. 2008. Induced pluripotent stem cells generated from patients with ALS can be differentiated into motor neurons. *Science* 321:1218–1221.

18. Ebert A. D., Yu J., Rose F. F. Jr, Mattis V. B., Lorson C. L., Thomson J. A., Svendsen CN. 2009. Induced pluripotent stem cells from a spinal muscular atrophy patient. *Nature* 457:277–280.

19. Eminli S., Foudi A., Stadtfeld M., Maherali N., Ahfeldt T., Mostoslavsky G., Hock H., Hochedlinger K. 2009. Differentiation stage determines potential of hematopoietic cells for reprogramming into induced pluripotent stem cells. *Nat Genet* 41:968–976.

20. Fraidenraich D., Benezra R. 2006. Embryonic stem cells prevent developmental cardiac defects in mice. *Nat Clin Pract Cardiovasc Med* 3 Suppl 1:S14–17.

21. French A. J., Adams CA, Anderson LS, Kitchen JR, Hughes MR, Wood SH. 2008. Development of human cloned blastocysts following somatic cell nuclear transfer with adult fibroblasts. *Stem Cells* 26:485–493.

22. Goldman S. 2005. Stem and progenitor cell-based therapy of the human central nervous system *Nat Biotechnol* 23:862–871.

23. Hanna J., Markoulaki S., Schorderet P., Carey BW, Beard C., Wernig M., Creyghton MP, Steine EJ, Cassady JP, Foreman

R., Lengner CJ, Dausman J. A., Jaenisch R. 2008. Direct reprogramming of terminally differentiated mature B lymphocytes to pluripotency. *Cell* 133:250–264.

24. Hanna J., Wernig M., Markoulaki S., Sun CW, Meissner A., Cassady JP, Beard C., Brambrink T., Wu LC, Townes TM, Jaenisch R. 2007. Treatment of sickle cell anemia mouse model with iPS cells generated from autologous skin. *Science* 318:1920–1923.

25. Hong H., Takahashi K., Ichisaka T., Aoi T., Kanagawa O, Nakagawa M., Okita K., Yamanaka S. 2009. Suppression of induced pluripotent stem cell generation by the p53-p21 pathway. *Nature* 460:1132–1135.

26. Hotta A., Ellis J. 2008. Retroviral vector silencing during iPS cell induction: an epigenetic beacon that signals distinct pluripotent states. *J Cell Biochem* 105:940–948.

27. Huangfu D., Maehr R., Guo W., Eijkelenboom A., Snitow M., Chen AE, Melton DA. 2008. Induction of pluripotent stem cells by defined factors is greatly improved by small-molecule compounds. *Nat Biotechnol* 26:795–797.

28. Jaenisch R., Young R. 2008. Stem cells, the molecular circuitry of pluripotency and nuclear reprogramming. *Cell* 132:567–582.

29. Kaji K., Norrby K., Paca A., Mileikovsky M., Mohseni P., Woltjen K. 2009. Virus-free induction of pluripotency and subsequent excision of reprogramming factors. *Nature* 458:771–775.

30. Karumbayaram S., Novitch B. G., Patterson M., Umbach J. A., Richter L., Lindgren A., Conway AE, Clark AT, Goldman SA, Plath K., Wiedau-Pazos M., Kornblum HI, Lowry WE. 2009.

Directed differentiation of human-induced pluripotent stem cells generates active motor neurons. *Stem Cells* 27:806–811.

31. Kawamura T., Suzuki J., Wang YV, Menendez S., Morera L. B., Raya A., Wahl GM, Belmonte J. C. 2009. Linking the p53 tumour suppressor pathway to somatic cell reprogramming. *Nature* 460:1140–1144.

32. Kim D., Kim C. H., Moon JI, Chung YG, Chang MY, Han BS, Ko S., Yang E., Cha KY, Lanza R., Kim KS. 2009. Generation of human induced pluripotent stem cells by direct delivery of reprogramming proteins. *Cell Stem Cell* 4:472–476.

33. Kim JB, Sebastiano V., Wu G., Araúzo-Bravo M. J., Sasse P., Gentile L., Ko K., Ruau D., Ehrich M., van den Boom D., Meyer J., Hübner K., Bernemann C., Ortmeier C., Zenke M., Fleischmann BK, Zaehres H., Schöler H. R. 2009. *Oct4*-induced pluripotency in adult neural stem cells. *Cell* 136:411–419.

34. Kim JB, Zaehres H., Wu G., Gentile L., Ko K., Sebastiano V., Araúzo-Bravo M. J., Ruau D., Han D. W., Zenke M., Schöler H. R. 2008. Pluripotent stem cells induced from adult neural stem cells by reprogramming with two factors. *Nature* 454:646–650.

35. Koch CA, Geraldes P., Platt J. L.. 2008. Immunosuppression by embryonic stem cells. *Stem Cells* 26:89–98.

36. Kroon E., Martinson LA, Kadoya K., Bang A. G., Kelly OG, Eliazer S., Young H., Richardson M., Smart NG, Cunningham J., Agulnick AD, D'Amour KA, Carpenter MK, Baetge EE. 2008. Pancreatic endoderm derived from human embryonic stem cells generates glucose-responsive insulin-secreting cells in vivo. *Nat Biotechno* 26:443–452.

37. Laflamme MA, Murry CE. 2005. Regenerating the heart. *Nat Biotechnol* 23:845–856.

38. Le Blanc K., Ringdén O. 2007. Immunomodulation by mesenchymal stem cells and clinical experience. *J Intern Med* 262:509–525.

39. Lee G., Papapetrou EP, Kim H., Chambers SM, Tomishima M. J., Fasano CA, Ganat YM, Menon J., Shimizu F., Viale A, Tabar V., Sadelain M., Studer L. 2009. Modelling pathogenesis and treatment of familial dysautonomia using patient-specific iPSCs. *Nature* 461:402–406.

40. Li H., Collado M., Villasante A., Strati K., Ortega S., Cañamero M., Blasco MA, Serrano M. 2009. The Ink4/Arf locus is a barrier for iPS cell reprogramming. *Nature* 460:1136–1139.

41. Lin T., Ambasudhan R., Yuan X., Li W., Hilcove S., Abujarour R., Lin X., Hahm HS, Hao E., Hayek A., Ding S. 2009. A chemical platform for improved induction of human iPSCs. *Nat Methods* 6:805–808.

42. Loh YH, Agarwal S., Park IH, Urbach A., Huo H., Heffner GC, Kim K., Miller JD, Ng K., Daley GQ. 2009. Generation of induced pluripotent stem cells from human blood. *Blood* 113:5476–5479.

43. Maehr R., Chen S., Snitow M., Ludwig T., Yagasaki L., Goland R., Leibel RL, Melton DA. 2009. Generation of pluripotent stem cells from patients with type 1 diabetes. *Proc Natl Acad Sci USA* 106:15768–15773.

44. Marchetto MC, Yeo GW, Kainohana O, Marsala M., Gage FH, Muotri AR. 2009. Transcriptional Signature and Memory Retention of Human-Induced Pluripotent Stem Cells. *PLoS ONE* 4:e7076.

45. Marión RM, Strati K., Li H., Murga M., Blanco R., Ortega S., Fernandez-Capetillo O, Serrano M., Blasco MA. 2009. A p53-mediated DNA damage response limits reprogramming to ensure iPS cell genomic integrity. *Nature* 460:1149–1153.

46. Martinez-Fernandez A., Nelson T. J., Yamada S., Reyes S., Alekseev AE, Perez-Terzic C., Ikeda Y., Terzic A. 2009. iPS Programmed without c-MYC yield proficient cardiogenesis for functional heart chimerism. *Circ Res* 105:648–656.

47. Mauritz C., Schwanke K., Reppel M., Neef S., Katsirntaki K., Maier LS, Nguemo F., Menke S., Haustein M., Hescheler J., Hasenfuss G., Martin U. 2008. Generation of functional murine cardiac myocytes from induced pluripotent stem cells. *Circulation* 118:507–517.

48. Meissner A., Wernig M., Jaenisch R. 2007. Direct reprogramming of genetically unmodified fibroblasts into pluripotent stem cells. *Nat Biotechnol* 25:1177–1181.

49. Meyer JS, Shearer RL, Capowski EE, Wright LS, Wallace KA, McMillan EL, Zhang SC, Gamm D. M. 2009. Modeling early retinal development with human embryonic and induced pluripotent stem cells. *Proc Natl Acad Sci USA* 106:16698–16703.

50. Mikkelsen TS, Hanna J., Zhang X., Ku M., Wernig M., Schorderet P., Bernstein BE, Jaenisch R., Lander ES, Meissner A. 2008. Dissecting direct reprogramming through integrative genomic analysis. *Nature* 454:49–55.

51. Miura K., Okada Y., Aoi T., Okada A., Takahashi K., Okita K., Nakagawa M., Koyanagi M., Tanabe K., Ohnuki M., Ogawa D., Ikeda E., Okano H., Yamanaka S. 2009. Variation in the safety of induced pluripotent stem cell lines. *Nat Biotechnol* 27:743–745.

52. Murry CE, Keller G. 2008. Differentiation of embryonic stem cells to clinically relevant populations: Lessons from embryonic development. *Cell* 132:661–680.

53. Nakagawa M., Koyanagi M., Tanabe K., Takahashi K., Ichisaka T., Aoi T., Okita K., Mochiduki Y., Takizawa N., Yamanaka S. 2008. Generation of induced pluripotent stem cells without MYC from mouse and human fibroblasts. *Nat Biotechnol* 26:101–106.

54. Narazaki G., Uosaki H., Teranishi M., Okita K., Kim B., Matsuoka S., Yamanaka S., Yamashita JK. 2008. Directed and systematic differentiation of cardiovascular cells from mouse induced pluripotent stem cells. *Circulation* 118:498–506.

55. Nelson T. J., Behfar A., Terzic A. 2008. Stem cells: biologics for regeneration. *Clin Pharmacol Ther* 84:620–623.

56. Nelson T. J., Behfar A., Terzic A. 2008. Strategies for Therapeutic Repair: The "R" Regenerative Medicine Paradigm. *Clin Transl Sci* 1:168–171.

57. Nelson T. J., Behfar A., Yamada S., Martinez-Fernandez A., Terzic A. 2009. Stem cell platforms for regenerative medicine. *Clin Trans Sci* 2:222–227.

58. Nelson T. J., Faustino R. S., Chiriac A., Crespo-Diaz R., Behfar A., Terzic A. 2008. CXCR4⁺/FLK-1⁺ biomarkers select a cardiopoietic lineage from embryonic stem cells. *Stem Cells* 26:1464–1473.

59. Nelson T. J., Martinez-Fernandez A., Terzic A. 2009. KCNJ11 knockout morula re-engineered by stem cell diploid aggregation. *Philos Trans R Soc Lond B Biol Sci* 364:269–276.

60. Nelson T. J., Martinez-Fernandez A., Yamada S., Mael AA, Terzic A., Ikeda Y. 2009. Induced pluripotent reprogramming from promiscuous human stemness-related factors. *Clin Transl Sci* 2:118–126.

61. Nelson T. J., Martinez-Fernandez A., Yamada S., Perez-Terzic C., Ikeda Y., Terzic A. 2009. Repair of acute myocardial infarction with human stemness factors induced pluirpotent stem cells. *Circulation* 120:408–416.

62. Nelson T. J., Martinez-Fernandez A., Yamada S., Terzic A. 2009. Induced pluripotent stem cells: Advances to applications. *Stem cells and Cloning* 0:000–000.

63. Nelson T. J., Terzic A. 2009. Induced pluripotent stem cells: reprogrammed without a trace. *Regen Med* 4:333–335.

64. Niwa A., Umeda K., Chang H., Saito M., Okita K., Takahashi K., Nakagawa M., Yamanaka S., Nakahata T., Heike T. 2009. Orderly hematopoietic development of induced pluripotent stem cells via Flk-1$^+$ hemoangiogenic progenitors. *J Cell Physiol* 221:367–77.

65. Okita K., Nakagawa M., Hyenjong H., Ichisaka T., Yamanaka S. 2008. Generation of mouse induced pluripotent stem cells without viral vectors. *Science* 322:949–953.

66. Okita K., Ichisaka T., Yamanaka S. 2007. Generation of germline-competent induced pluripotent stem cells. *Nature* 448:313–317.

67. Orkin SH, Zon LI. 2008. Hematopoiesis: An evolving paradigm for stem cell biology. *Cell* 132:631–644.

68. Osakada F., Jin ZB, Hirami Y., Ikeda H., Danjyo T., Watanabe K., Sasai Y., Takahashi M. 2009. In vitro differentiation of retinal

cells from human pluripotent stem cells by small-molecule induction. *J Cell Sci* 122:3169–3179.

69. Park IH, Arora N., Huo H., Maherali N., Ahfeldt T., Shimamura A., Lensch MW, Cowan C., Hochedlinger K., Daley GQ. 2008. Disease-specific induced pluripotent stem cells. *Cell* 134:877–886.

70. Phinney DG, Prockop DJ. 2007. Mesenchymal stem/multipotent stromal cells: the state of transdifferentiation and modes of tissue repair. *Stem Cells* 25:2896–2902.

71. Quaini F., Urbanek K., Beltrami AP, Finato N., Beltrami CA, Nadal-Ginard B., Kajstura J., Leri A., Anversa P. 2002. Chimerism of the transplanted heart. *N Engl J Med* 346:5–15.

72. Raya A., Rodríguez-Pizà I., Guenechea G., Vassena R., Navarro S., Barrero M. J., Consiglio A., Castellà M., Río P., Sleep E., González F., Tiscornia G., Garreta E., Aasen T., Veiga A., Verma IM, Surrallés J., Bueren J., Izpisúa Belmonte J. C. 2009. Disease-corrected haematopoietic progenitors from Fanconi anaemia induced pluripotent stem cells. *Nature* 460:53–59.

73. Schenke-Layland K., Rhodes KE, Angelis E., Butylkova Y., Heydarkhan-Hagvall S., Gekas C., Zhang R., Goldhaber JI, Mikkola HK, Plath K., MacLellan WR. 2008. Reprogrammed mouse fibroblasts differentiate into cells of the cardiovascular and hematopoietic lineages. *Stem Cells* 26:1537–1546.

74. Senju S., Haruta M., Matsunaga Y., Fukushima S., Ikeda T., Takahashi K., Okita K., Yamanaka S., Nishimura Y. 2009. Characterization of dendritic cells and macrophages generated by directed differentiation from mouse induced pluripotent stem cells. *Stem Cells* 27:1021–1031.

75. Silva J., Barrandon O, Nichols J., Kawaguchi J., Theunissen TW, Smith A. 2008. Promotion of reprogramming to ground state pluripotency by signal inhibition. *PLoS Biol* 6:e253.

76. Silva J., Smith A. 2008. Capturing pluripotency. *Cell* 132:532–536.

77. Si-Tayeb K., Noto FK, Nagaoka M., Li J., Battle MA, Duris C., North PE, Dalton S., Duncan SA. 2009. Highly efficient generation of human hepatocyte-like cells from induced pluripotent stem cells. *Hepatology* 0:000–000, DOI:10.1002/hep.23354.

78. Solter D. 2006. From teratocarcinomas to embryonic stem cells and beyond: a history of embryonic stem cell research. *Nat Rev Genet* 7:319–327.

79. Song Z., Cai J., Liu Y., Zhao D., Yong J., Duo S., Song X., Guo Y., Zhao Y., Qin H., Yin X., Wu C., Che J., Lu S., Ding M., Deng H. 2009. Efficient generation of hepatocyte-like cells from human induced pluripotent stem cells. *Cell Res* 19:1233–1242.

80. Stadtfeld M., Brennand K., Hochedlinger K. 2008. Reprogramming of pancreatic beta cells into induced pluripotent stem cells. *Curr Biol* 18:890–894.

81. Stadtfeld M., Nagaya M., Utikal J., Weir G., Hochedlinger K. 2008. Induced pluripotent stem cells generated without viral integration. *Science* 322:945–994.

82. Sun N., Panetta NJ, Gupta D. M., Wilson KD, Lee A., Jia F., Hu S., Cherry AM, Robbins RC, Longaker MT, Wu J. C. 2009. Feeder-free derivation of induced pluripotent stem cells from adult human adipose stem cells. *Proc Natl Acad Sci USA* 106:15720–15725.

83. Tachibana M., Sparman M., Sritanaudomchai H., Ma H., Clepper L., Woodward J., Li Y., Ramsey C., Kolotushkina O, Mitalipov S. 2009. Mitochondrial gene replacement in primate offspring and embryonic stem cells. *Nature* 461:367–372.

84. Takahashi K., Tanabe K., Ohnuki M., Narita M., Ichisaka T., Tomoda K., Yamanaka S. 2007. Induction of pluripotent stem cells from adult human fibroblasts by defined factors. *Cell* 131:861–872.

85. Takahashi K., Yamanaka S. 2006. Induction of pluripotent stem cells from mouse embryonic and adult fibroblast cultures by defined factors. *Cell* 126:663–676.

86. Tashiro K., Inamura M., Kawabata K., Sakurai F., Yamanishi K., Hayakawa T., Mizuguchi H. 2009. Efficient adipocyte and osteoblast differentiation from mouse induced pluripotent stem cells by adenoviral transduction. *Stem Cells* 27:1802–1811.

87. Tateishi K., He J., Taranova O, Liang G., D.'Alessio AC, Zhang Y. 2008. Generation of insulin-secreting islet-like clusters from human skin fibroblasts. *J Biol Chem* 283:31601–31607.

88. Terzic A., Nelson T. 2010. Regenerative medicine: Advancing healthcare 2020. *J Am Coll Cardiol* 0:000–000.

89. Thomson J. A., Itskovitz-Eldor J., Shapiro SS, Waknitz MA, Swiergiel J. J., Marshall VS, Jones JM. 1998. Embryonic stem cell lines derived from human blastocysts. *Science* 282:1145–1147.

90. Utikal J., Polo JM, Stadtfeld M., Maherali N., Kulalert W., Walsh RM, Khalil A., Rheinwald JG, Hochedlinger K. 2009.

Immortalization eliminates a roadblock during cellular re-programming into iPS cells. *Nature* 460:1145–1148.

91. Van de Ven C., Collins D., Bradley MB, Morris E., Cairo MS. 2007. The potential of umbilical cord blood multipotent stem cells for nonhematopoietic tissue and cell regeneration. *Exp Hematol* 35:1753–1765.

92. Waldman SA, Terzic A. 2007 .Individualized medicine and the imperative of global health. *Clin Pharmacol Ther* 82:479–483.

93. Wagers A. J., Weissman IL. 2004. Plasticity of adult stem cells. *Cell* 116:639–648.

94. Wernig M., Zhao JP, Pruszak J., Hedlund E., Fu D., Soldner F., Broccoli V., Constantine-Paton M., Isacson O, Jaenisch R. 2008. Neurons derived from reprogrammed fibroblasts functionally integrate into the fetal brain and improve symptoms of rats with Parkinson's disease. *Proc Natl Acad Sci USA* 105:5856–5861.

95. Woltjen K., Michael IP, Mohseni P., Desai R., Mileikovsky M., Hämäläinen R., Cowling R., Wang W., Liu P., Gertsenstein M., Kaji K., Sung HK, Nagy A. 2009. piggyBac transposition reprograms fibroblasts to induced pluripotent stem cells. *Nature* 458:766–770.

96. Xu D., Alipio Z., Fink LM, Adcock D. M., Yang J., Ward DC, Ma Y. 2009. Phenotypic correction of murine hemophilia A using an iPS cell-based therapy. *Proc Natl Acad Sci USA* 106:808–813.

97. Yamada S., Nelson T. J., Crespo-Diaz RJ, Perez-Terzic C., Liu XK, Miki T., Seino S., Behfar A., Terzic A. 2008. Embryonic stem cell therapy of heart failure in genetic cardiomyopathy. *Stem Cells* 26:2644–2653.

98. Yamada S., Nelson T. J., Behfar A., Crespo-Diaz RJ, Fraiden-raich D., Terzic A. 2009. Stem cell transplant into preimplan-tation embryo yields myocardial infarction-resistant adult phenotype. *Stem Cells* 27:1697–1705.

99. Yamanaka, S. 2009. Elite and stochastic models for induced pluripotent stem cell generation. *Nature* 460:49–52.

100. Yang X., Smith S. L., Tian XC, Lewin HA, Renard JP, Wakayama T. 2007. Nuclear reprogramming of cloned embryos and its implications for therapeutic cloning. *Nat Genet* 39:295–302.

101. Ye Z., Zhan H., Mali P., Dowey S., Williams D. M., Jang YY, Dang CV, Spivak J. L., Moliterno AR, Cheng L. 2009. Human induced pluripotent stem cells from blood cells of healthy donors and patients with acquired blood disorders. *Blood* 0:000–000.

102. Yokoo N., Baba S., Kaichi S., Niwa A., Mima T., Doi H., Ya-manaka S., Nakahata T., Heike T. 2009. The effects of car-dioactive drugs on cardiomyocytes derived from human induced pluripotent stem cells. *Biochem Biophys Res Com-mun* 387:482–488.

103. Yoshida Y., Takahashi K., Okita K., Ichisaka T., Yamanaka S. 2009. Hypoxia enhances the generation of induced pluripo-tent stem cells. *Cell Stem Cell* 5:237–241.

104. Yu J., Hu K., Smuga-Otto K., Tian S., Stewart R., Slukvin II, Thom-son J. A. 2009. Human Induced Pluripotent Stem Cells Free of Vector and Transgene Sequences. *Science* 324:797–801.

105. Yu J., Vodyanik MA, Smuga-Otto K., Antosiewicz-Bourget J., Frane J. L., Tian S., Nie J., Jonsdottir GA, Ruotti V., Stew-art R., Slukvin II, Thomson J. A. 2007. Induced pluripotent

stem cell lines derived from human somatic cells. *Science* 318:1917–1920.

106. Zhao XY, Li W., Lv Z., Liu L., Tong M., Hai T., Hao J., Guo CL, Ma QW, Wang L., Zeng F., Zhou Q. 2009. iPS cells produce viable mice through tetraploid complementation. *Nature* 461:86–90.

107. Zhang D., Jiang W., Liu M., Sui X., Yin X., Chen S., Shi Y., Deng H. 2009. Highly efficient differentiation of human ES cells and iPS cells into mature pancreatic insulin-producing cells. *Cell Res* Cell Res 19:429–438.

108. Zhang J., Wilson G. F., Soerens A. G., Koonce C. H., Yu J., Palecek S. P., Thomson J. A., Kamp T. J. 2009. Functional cardiomyocytes derived from human induced pluripotent stem cells. *Circ Res* 104:e30–e41.

109. Zhou H., Wu S., Joo J. Y., Zhu S., Han D. W., Lin T., Trauger S., Bien G., Yao S., Zhu Y., Siuzdak G., Schöler H. R., Duan L., Ding S. 2009. Generation of induced pluripotent stem cells using recombinant proteins. *Cell Stem Cell* 4:381–384.

Chapter 11

Strategies to Fabricate Engineered Cardiac Tissues

Paolo Di Nardo[1,3,5], Giancarlo Forte[1,2], Marilena Minieri[1,3,4], Luciana Carosella[4,5]

[1]Laboratorio di Cardiologia Molecolare e Cellulare, Dipartimento di Medicina Interna, Università di Roma Tor Vergata, Roma, Italy
[2]Biomaterials Center, International Center for Materials Nano-architectonics (MANA), National Institute for Materials Science (NIMS), Tsukuba, Japan
[3]Japanese-Italian Tissue Engineering Laboratory (JITEL), Tokyo Women's Medical University-Waseda University Joint Institution for Advanced Biomedical Sciences (TWIns), Tokyo, Japan
[4]Canadian-Italian Tissue Engineering Laboratory (CITEL), St. Boniface Research Center, Winnipeg, Manitoba, Canada
[5]BioLink Institute, Link Campus University, Rome, Italy
[6]Istituto di Medicina Interna e Geriatria, Università Cattolica S. Cuore, Roma, Italy

For correspondence:
Paolo Di Nardo, MD
Dipartimento di Medicina Interna
Università di Roma Tor Vergata
Via Montpellier, 1
00133 Roma, Italy

E-mail: dinardo@uniroma2.it
Phone: +39-06-72594215
Fax: +39-06-2024130 or +39-06-72594263

1. Introduction

A decade of intensive scientific and economic efforts to implement cardiac cell therapy has demonstrated that the approaches so far proposed are unfit to allow clinically safe, reliable, and cost-effective procedures. De facto, stem cell manipulation is a very complex endeavor for which a completely innovative vision and novel technologies are necessary. Instead, the overwhelming first wave of enthusiasm about the possibility of rejuvenating biological tissues and organs has suggested to rapidly adopt simplistic solutions to technologically exploit the stem cell potential. The inappropriateness of current cell therapy protocols is testified by the meager number (approx. 3–10%) of stem cells homing after injection into the myocardium independently of the approach adopted (intramyocardial, intracoronary, retrograde coronary venous). In fact, most of the injected cells do not home into the heart, but are suppressed by apoptosis or removed by the bloodstream and, then, entrapped into the lungs [1].

The failure to obtain a suitable myocardial repair by injecting isolated stem cells can be attributed to drawbacks in, at least, two out of four steps in which the cardiac cell therapy process is articulated. In fact, in a first step, few stem cells are isolated from a donor and purified by the expression of a membrane protein complex representative of the cell phenotype. In a second step, isolated stem cells are expanded in vitro to generate a population sufficiently large to match, after delivery, the needs of the recipient organ (third step). Finally, in the fourth step, stem cells home into the damaged myocardium. In this process, the second (cell expansion) and third step (cell delivery) are very critical. In fact, based on current protocols, just isolated stem cells are cultured for expansion and then detached from the support using procedures established decades ago for differentiated cells. These procedures neglect that the stem cells unique characteristics can be preserved in vitro only if cells are cultured in a very special environment. In fact, in vivo, stem cells are not intermixed with differentiated cells, but segregated in special regions (niches), where the micro-environment contributes to preserve their peculiar characteristics and from which they migrate and differentiate upon appropriate stimuli to restore the damaged tissue [2]. Therefore, after isolation, the stem cells phenotypic integrity can be maintained in vitro only if environmental conditions emulating the native niche are reproduced. Instead, stem cells improperly cultured in conventional culture conditions are exposed to enzyme manipulation that damages cell membrane and destroys the self-produced extracellular matrix required for progenitor cell homing and differentiation [3,4]. Cells deprived of their natural adhesion structures are subsequently delivered as single cell suspension. The delivery system so far adopted, injection that does not require a specific surgical skill, is inexpensive and can be used in most organs. Unfortunately, injection exposes suboptimal suspended stem cells to additional distressing conditions (such as syringe and intra-needle pressure, and host's critical ischemic microenvironment), to which they, usually embedded in the less-stressed atrial or apical myocardium [5], are not structurally and functionally adapted. As a result, few injected cells home and differentiate into the myocardium with inconsistent advantages for the cardiac tissue structure and function [6]. Indeed, after injection, the implanted stem cell number and differentiation as well as the graft shape, size, and location are unpredictable. The maximization of the indubitable repairing potential of stem cells requires that the knowledge of the stem cell behaviour is preliminarily improved, and adequate procedures are setup to easily govern cell fate.

In this context, cell injection could be more efficiently replaced by more sophisticated procedures inspired by tissue engineering concepts. Combining cells and scaffolds made of biocompatible materials, it would be possible to fabricate ex vivo portions of tissue to be re-implanted into the damaged organs. This approach requires a multilevel strategy (figure 1) in which cells to be implanted (not necessarily stem cells) are carefully studied, and scaffolds appropriately designed. In a further step, procedures to match cells and scaffolds in order to fabricate engineered tissues as well as to deliver them must be setup. Finally, protocols to induce stem cells embedded into the engineered tissues to differentiate must be carefully defined. In this context, differentiation cannot represent the final goal. In fact, generating perfectly differentiated cells without strategies to integrate them in the native tissue architecture is ineffectual in respect to the organ's structural and functional rejuvenation. Therefore, the ideal road map to develop clinically efficient engineered tissues should include protocols to jointly induce stem cell differentiation and integration in the recipient tissue.

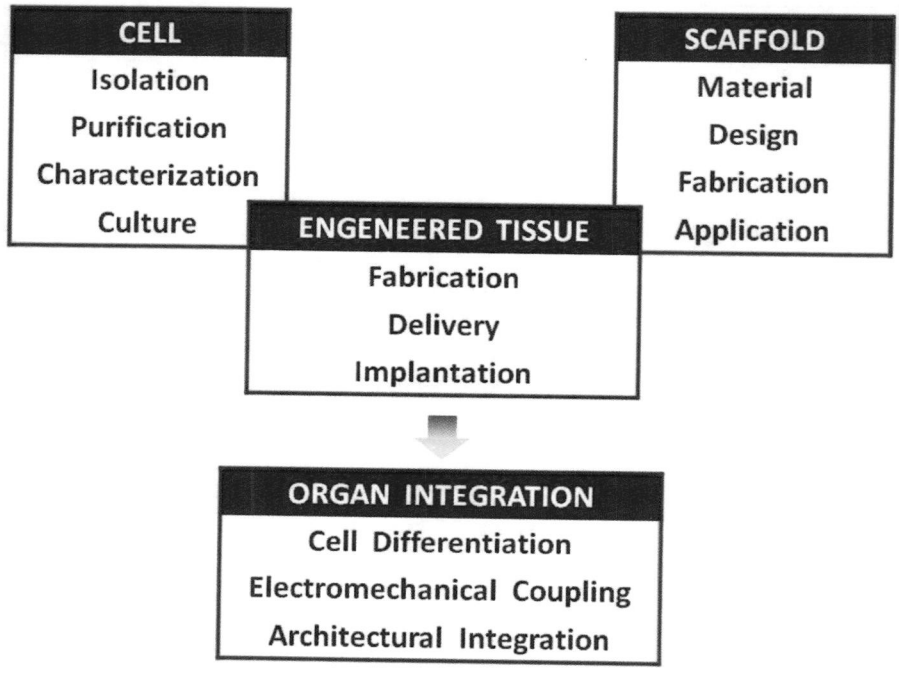

Figure 1. Tissue engineering requires a multilevel approach and strictly standardized procedures in each step. Governing stem cell differentiation and integration into the host organ is crucial.

2. Selecting Cells for Cardiac Tissue Engineering

In early investigations to repair the myocardial damage, the inability of postnatal cardiomyocytes to proliferate [7] suggested exploiting the skeletal myoblasts' capability to express a contractile phenotype in vitro [8] in order to generate new cardiomyocytes in vivo. Indeed, after implantation in canine hearts affected by dilated cardiomyopathy (DCM), skeletal myoblasts reduced cardiac remodeling [9], but the effects were more related to their fusion with the surrounding myocardial cells than to direct cell differentiation [10]. Finally, clinical trials demonstrated that skeletal myoblasts were not able to electrically couple with cells in the host tissue, inducing severe arrhythmic events [11]. These frustrating results induced researchers to pay more attention to the possible use of stem/progenitor cells for fabricating engineered cardiac tissues. Stem cells are present in the native tissues in a very limited number and require complex protocols to be expanded to obtain a quantity sufficient for tissue fabrication. During the expansion process in conventional culture conditions, stem cells can undergo senescence after a few passages in vitro [12] or face malignant transformation [13,14]. Additionally, stem cells can suffer from a lot-to-lot variability in quality [15], and the use of animal-derived supplements could be responsible of immune rejection events. Finally, treatments of human diseases based on cellular and tissue products are subjected to stringent regulations by the European Union (EU) and the U.S. Food and Drug Administration (FDA). In particular, the EU requires that Good Manufacturing Practice (GMP) protocols must be adopted for cell therapy in human beings [16,17]. Taken together, the use of stem cells for the treatment of human diseases is still affected by major impediments, most of which are caused by the lack of standardization about many technical and regulatory aspects. Stem cell technology is among the most complex research endeavors and needs materials and procedures to be strictly standardized through a long-term process actuated by merging the quantum of knowledge resident in different disciplines and international laboratories, avoiding fragmented knowledge and unsuitable options for innovative treatments to be delivered in the clinical setting. Standardized procedures are particularly critical to select the most appropriate cell subset to generate new vessels and contractile cardiomyocytes. Ideally, embryonic stem cells could represent the optimal choice for generating new cardiomyocytes to be implanted in injured hearts. However, ethical issues and the current level of knowledge and technology related to embryonic stem cells do not allow consideration of their therapeutic use at hand. In addition, it must be taken in due consideration that embryonic stem cells do not exclude the host immunoreaction, the transfer of potentially noxious agents to the recipient, and the potential transformation in neoplastic cells. More realistic, at present, could be to use adult stem cells that can be easily isolated from different patients' tissues [18] and expanded in culture. Furthermore, the same patient could act as donor and recipient of the adult stem cell implant. Among the adult stem cells are the resident cardiac progenitor cells (CPC), originally believed to originate from the bone-marrow through the bloodstream owing to the apparent commonality of several antigenic stemness markers. Indeed, bone marrow–derived stem cells are only minimally able to generate new contractile cells, while a paracrine effect on the diseased myocardial tissue is universally recognized [19]. They also induce a certain degree of immune tolerance [20]. Actually, resident CPC are leftovers of the embryonic life entrapped in the differentiated mammalian tissue [21], where, in strict combination with supporting cells, they reside within special regions (niches) and remain quiescent for short or long periods of time [22].

CPC were initially identified as stem-like cells within rodent, canine, and human myocardium because they express the *c-kit* (*CD117*) antigen [5]. These cells, also expressing *multidrug resistance-1* (*MDR-1*) and *Stem cell antigen-1* (*Sca-1*), are self-renewing, clonogenic, and able to differentiate in vitro into the three principal cardiac cell types (CCT): cardiomyocytes, smooth muscle, and endothelial cells. However, the experimental disruption of the *Kit* gene in mice mainly affected marrow-derived hematopoietic and endothelial cell development reducing progenitor-cell mobilization from marrow and the release of cytokines and chemokines that may participate in the cardioprotective paracrine signaling [23,24]. Furthermore, many laboratories failed to isolate *c-kit*pos cells and thus to confirm initial results. More likely, *c-kit* only transiently marks cardiac progenitors, which, in rodents [2] and human beings, are more stably identified by the expression of *Sca-1*, an antigen originally identified in activated lymphocytes. In respect to *c-kit*pos cells, *Sca-1*pos cells are relatively abundant in fetal and adult hearts and, thus, more easily isolable from human biopsies and expandable in culture. Different investigations have extensively demonstrated murine and human *Sca-1*pos CPC multipotency and capability to differentiate to cardiomyocytes [25, 26, 27] when properly stimulated. Indeed, the assessment of the multipotency is usually performed in conventional culture conditions, without considering the peculiar characteristics of the stem cell behavior and environment. Recently, it has been suggested that at least three classes of factors (physico-chemical, biochemical, and mechanostructural) contribute to determine the stem cell fate [3,4] (figure 2). The differentiation toward a specific phenotype could be obtained only if stem cells are cultured in an environment mimicking the native cell conditions and characterized by the optimal symmetry among different factors partaking to the above mentioned classes. Different levels of symmetry will drive the achievement of a specific phenotype for a long time. In this context, a properly designed scaffold plays a pivotal role as the source of the mechano-structural signals, while the physico-chemical and biochemical factors can be easily regulated through the culture medium.

313

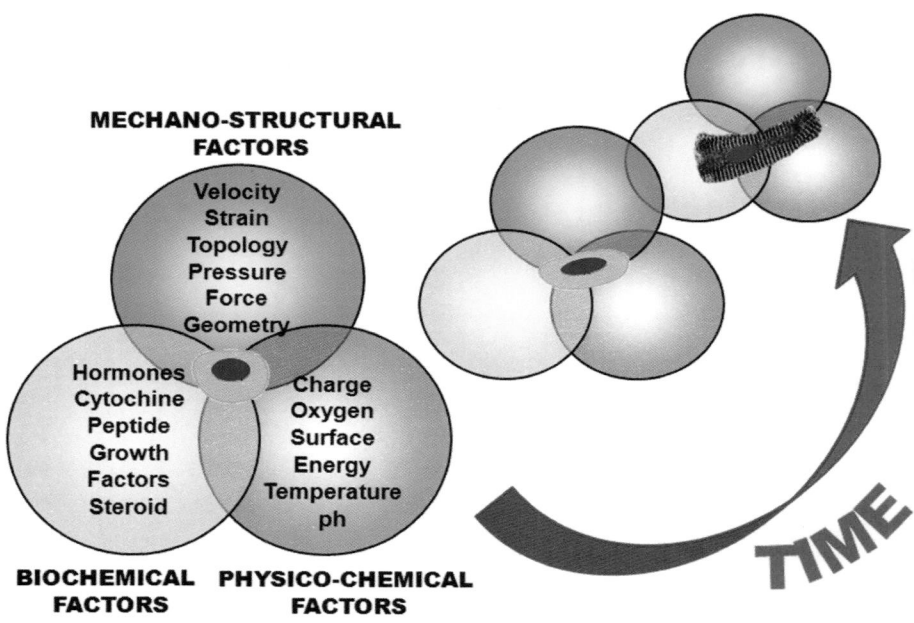

Figure 2. The three classes of stimuli, which interact dynamically to modulate the symmetric environment throughout which stem cells differentiate. Note that time induces the features of the symmetric environment to change.

3. Designing Inherently Bio-Active Scaffolds

Numerous experiments and options have since been evaluated in order to identify the best materials and the optimum technologies to design and process scaffolds for cardiac tissue engineering. Based on current concepts, scaffolds should display a series of characteristics (table 1) to suitably correspond to clinical requirements and allow the formation of an engineered tissue with predetermined shape. However, a consistent design strategy and well-founded engineering principles and models have been sorely missing from the research scenario. Most current

314

scaffolds are made of biologically inert biocompatible materials and are intended as mere mechanical support for cells. In order to supply seeded cells with specific biological signals, processes to fabricate smart scaffolds, in which biologically active molecules (growth factors, cytokines, etc.) are embedded into the inert material, are under scrutiny. However, there is no doubt that current materials and methods cannot deliver functional heart tissue, as required for cardiac regeneration, and a change in perspective and strategy is required. Consistently, very recently, it has been demonstrated that inert scaffolds can release signals sensed as biologically relevant by cells when the chemical structure and topology have been modified. This new class of scaffolds has been named Inherently Bio-Active Scaffolds (IBAS) [3]. In fact, the notion that cells are exclusively sensing biological signals released from remote tissues (hormones), extracellular matrix, and neighboring cells (growth factors, cytokines, adhesion factors, etc.) has been recently challenged by the evidence that the chemical, physical, and mechanical features (stiffness, surface roughness, porosity, micro- and nano-architecture, etc.) of the substrate can drive stem cell differentiation, even in the absence of specific biological cues [28, 29, 30]. Indeed, stem cells can be pre-committed toward a defined phenotype matching biological signals with physical and mechanical signals arising from the extracellular matrix [31–35]. Therefore, scaffolds used in tissue engineering cannot be a mere mechanical cell support, but must be able to release signals perceived as biologically relevant by both stem and differentiated cells, thus mimicking the in vivo microenvironment [36,37]. The role of the scaffold design is particularly evident when stem/progenitor cells are used to fabricate architecturally complex engineered tissues, such as the myocardium. The proper combination in vitro of physico-chemical and biological stimuli provided by polymeric scaffolds and neonatal cardiomyocytes, respectively, must emulate an artificial niche able to efficiently drive cardiac progenitor cell fate determination [3]. Indeed, scaffolds tessellated with appropriate pore geometry and distribution display stiffness tightly correlated with the phenotype and function of contractile cells [38]. Consistently, embryonic and neonatal cardiomyocytes quickly lose morphological and functional properties when cultured on rigid matrices [39,40]. These findings, although preliminary, demonstrate that the design of biologically active scaffolds for tissue regeneration is critically dependent on the fundamental understanding of how cells coordinate their functions in the in vivo environment and in engineered matrices. Therefore matrix-cell and scaffold-cell interactions must first be properly investigated through both numerical modeling and experimental assessment, and only then can they be validated in vitro and ultimately in vivo.

4. Fabricating Cardiac Engineered Tissues

Several protocols using differentiated and non-differentiated cells and diversely designed scaffolds have been proposed by investigators [41–45] to fabricate portions of mammalian heart tissue, but safe, efficient, and cost/effective clinical applications are not available yet. However, all these attempts were biased by some drawbacks. First, cells were cultured using conventional procedures unfit for maximizing their homing potential. Second, the target tissue in the host heart is subjected to inflammatory processes of different severity, and no specific investigation has been so far designed to define the optimal implantation timing or to develop procedures specifically designed to prepare the host tissue to receive the implant.

Finally, and most importantly, all experiments have been exclusively focused on fabricating heart portions made of a single cell type, that is, cardiomyocytes or stem cells supposed to differentiate to cardiomyocytes, neglecting the cell type multiplicity and architectural complexity of the heart. Indeed, cell types suitable for heart regeneration in humans wait to be exhaustively defined [46] and generating new cardiomyocytes may not be sufficient to efficiently repair the texture of the myocardial tissue, which also includes fibroblasts, smooth muscle cells, endothelial cells, and adipocytes, among others. It is worth noting that only 10–20% of the cell complex constituting the whole healthy human heart are cardiomyocytes [47,48]. Current technologies for heart repair have not been set up to mimic the native myocardial architecture, but have been formulated assuming that cell therapy of injured hearts can be limited only to restoring the original number of working cardiomyocytes. This approach failed to re-establish the organ integrity, and the modest structural and functional heart improvements reported by some investigators were due to paracrine factors rather than to the mechanical support of newly implanted cells [49]. Furthermore, attention must be concentrated on the possibility that the preservation of the functional efficiency in the injured hearts may not necessarily be related to the re-establishment of the original cardiomyocyte number. Patients with limited necrotic regions subsequent to heart infarction and adequate ejection fraction (>40) can survive without major restrictions to their daily activities. Consistently, the goal of the cardiac cell therapy should be to bring the myocardial cell number and the organ function above a critical threshold without necessarily substituting all the damaged cells. The implanted cells must differentiate to cardiomyocytes and other cell types able to integrate in different cardiac structures (muscle, vessels), contributing to attain an adequate heart function not only when localized damages are present within the myocardium (infarction), but also when injured cells are widespread throughout the entire cardiac tissue (cardiomyopathy). Therefore, current concepts to fabricate heart tissue portions must be revised, and novel strategies must be formulated in order to imitate the texture of the healthy myocardium. The reorientation of the research activity about cardiac tissue engineering will require the invention of novel materials and processes to develop a new generation of scaffolds and more accurate culturing protocols for cardiogenic progenitor cells.

5. Alternative Strategies for Heart Repair

Despite decades of experimental research on biomaterials and scaffolds, neither the ideal material nor architecture for cardiac cell proliferation and colonization has been identified. Moreover, the understanding of cardiomyocyte origin and potency is limited; there is a huge amount of controversy and misinformation regarding cardiac stem cell characteristics and their isolation, identification, and expansion. Therefore, relevant resources and extensive investigations are still necessary to translate the concepts of cardiac tissue engineering in clinical applications. This necessitates the exploration of novel alternative strategies for heart mending. These strategies must be aimed at avoiding the drawbacks so far experienced with conventional techniques. In other words, new protocols must be set up to culture stem cells imitating the intra-niche conditions and to fairly deliver cells into the myocardium. In this respect, among the most promising techniques is the attempt to eliminate the presence of the scaffold in the thickness of the engineered tissue. The scaffold is usually positioned inside (endo-scaffold) the engineered tissue for a predetermined period of time during which it is exposed to erosion by cells and environmental factors with consequent release of potentially detrimental by-products. To avoid the release of possibly harmful by-products one option could be to position the scaffold outside (exo-scaffold) the newly fabricated tissue. This approach is based on the use of temperature-responsive poly-N-isopropyl acrilamyde–coated (PnIPAAm) surfaces, on which cells are grown and easily detached as cell sheets by lowering the temperature. PNIPAAm can be dissolved in aqueous solution, adsorbed or grafted on aqueous-solid interfaces, or cross-linked as hydrogels. When heated in water above 32°C, it undergoes a reversible phase transition from a swollen hydrated state to a shrunken dehydrated state, losing about 90% of its mass [50,51]. Cells grown on the thermo-responsive surfaces establish consistent inter-cellular contacts and generate a tissue sheet that can be easily detached from the PIPAAm-grafted surface by reducing the temperature below 32°C. Recently, applying this technology, it has been possible to fabricate engineered cardiac tissues using human Sca-1^{pos} CPC. This newly-generated tissue displayed a remarkable capability, when leaned on the murine visceral pericardium, to allow many stem cells to migrate from the cell sheet to the myocardial tissue (figure 3). Migrated cells differentiated to cardiomyocytes and fully integrated in the architecture of the recipient myocardium. In addition, several CPC migrated to the coronary artery wall where they differentiated to smooth muscle cells and perfectly integrated in the vessel wall [52]. These findings unveiled that cell sheets, besides being a patch to mechanically support the damaged myocardium, could represent a sort of "artificial niche" in which stem cells grow in the absence of apoptosis and can preserve their self-produced extracellular matrix that plays a fundamental role in cell migration to the host tissue [53]. No other previously scrutinized delivery system demonstrated the capability of

319

preserving stem cells in their optimal status or the ability to induce a large number of cells to migrate and integrate into the host tissue. This technology holds promise to circumvent many of the drawbacks so far experienced with other delivery systems. However, a future challenge will be to merge cell sheet and the IBAS technology in order to refurbish the injured heart with adequate and properly integrated contractile and non-contractile cells. This will represent a big stride in the clinical application of tissue engineering. In fact, the knowledge and technology related to materials interfacing with the human body in any context—be it chemical sensors for biological analytes, systems for drug testing, or biomaterials—have made remarkably little progress in comparison with other technological domains (electronics, etc.). The list of materials approved for in vivo use has remained practically unchanged for decades. Biodegradable polyesters continue to be proposed for tissue engineering applications despite reports on inflammatory response and mechanical failure simply because no viable alternatives are available. Indeed, the key factor that must be addressed is the complexity of the human body and the interactions between different organs, systems, and tissues, as well as the profound role the environment and external chemical or physical factors have on the body's response. No material in the human body is monocomponent and all senses—be they chemical or physical—are integrated, processed, and fuzzy. The complexity must be challenged and met with complex materials and sensing systems that exploit the plethora of knowledge acquired in chemistry and physics to create new hybrid composite biomaterials and alternative transduction modalities. Like all modern science, that of materials and sensing is limited by its introspection and reductionism. New approaches should consider the mechanical, chemical, and topological signals and design new complex multi-shell or nanocomposite materials able to encapsulate cells, which not only allow induction or control of cell fate but also enable dynamic re-modeling of the ECM. Using intuitive and adaptive complex materials, different microsystems can be assembled into a modular engineered tissue.

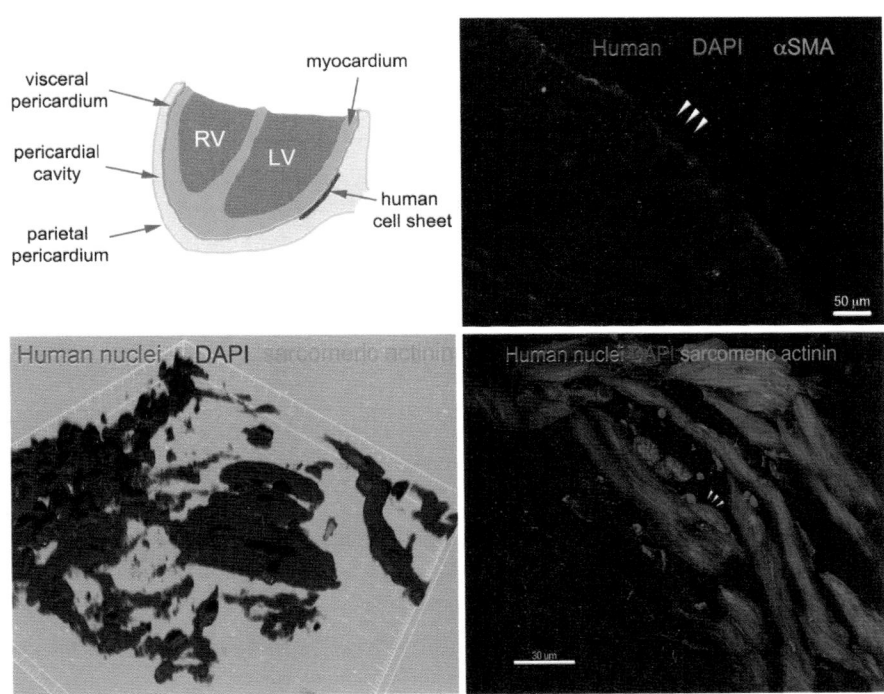

Figure 3. Human cardiac progenitor cell sheet leaned onto murine left ventricle. Murine parietal pericardium was removed, and human cell sheet leaned on the visceral pericardium (upper left). One week after the transplant, human cells could still be found attached to the heart (in red, upper right), and some human cells (highlighted by a red spot in the nucleus) could be found in contact with murine cardiomyocytes (bottom left) or expressing sarcomeric actinin (bottom right).

References

1.   Bui QT, Gertz ZM, Wilensky RL. 2010. Intracoronary delivery of bone-marrow-derived stem cells. *Stem Cell Res Ther* 1:29–35.

2.   Anversa P, Kajstura J, Leri A, Bolli R. 2006. Life and death of cardiac stem cells: A paradigm shift in cardiac biology. *Circulation* 113:1451–63.

3.   Pagliari S, Vilela-Silva A, Forte G, Pagliari F, Mandoli C, Vozzi G, Pietronave S, Prat M, Licoccia S, Ahluwalia A, Traversa E, Minieri M, Di Nardo P. 2011. Cooperation of biological and mechanical signals in cardiac progenitor cell differentiation. *Adv Mater* 23:514–18.

4.   Di Nardo P, Forte G, Ahluwalia A, Minieri M. 2010. Cardiac progenitor cells: potency and control. *J Cell Physiol* 224:590–600.

5.   Quaini F, Urbanek K, Beltrami AP, Finato N, Beltrami CA, Nadal-Ginard B, Kajstura J, Leri A, Anversa P. 2002. Chimerism of the transplanted heart. *N Engl J Med* 346:5–15.

6.   Müller-Ehmsen J, Krausgrill B, Burst V, Schenk K, Neisen UC, Fries JW, Fleischmann BK, Hescheler J, Schwinger RH. 2006. Effective engraftment but poor mid-term persistence of mononuclear and mesenchymal bone marrow cells in acute and chronic rat myocardial infarction. *J Mol Cell Cardiol* 41:876–84.

7.   Bergmann O, Bhardwaj RD, Bernard S, Zdunek S, Barnabé-Heider F, Walsh S, Zupicich J, Alkass K, Buchholz BA, Druid H, Jovinge S, Frisén J. 2009. Evidence for Cardiomyocyte Renewal in Humans. *Science* 324:98–102.

8.   Formigli L, Francini F, Tani A, Squecco R, Nosi D, Polidori L, Nis-
      tri S, Chiappini L, Cesati V, Pacini A, Perna AM, Orlandini GE,
      Zecchi Orlandini S, Bani D. 2005. Morphofunctional integra-
      tion between skeletal myoblasts and adult cardiomyocytes
      in coculture is favoured by direct cell-cell contacts and re-
      laxin treatment. *Am J Physiol Cell Physiol* 288:C795–804.

9.   Hata H, Matsumiya G, Miyagawa S, Kondoh H, Kawaguchi
      N, Matsuura N, Shimizu T, Okano T, Matsuda H, Sawa H. 2009.
      Grafted skeletal myoblasts sheets attenuate myocardial re-
      modelling in pacing-induced canine heart failure model. *J
      Thorac Cardiovasc Surg* 138:460–67.

10.   Reinecke H, Minami E, Poppa V, Murry CE. 2004. Evidence
      for fusion between cardiac and skeletal muscle cells. *Circ
      Res* 94:e56–e60.

11.   Menasché P, Alfieri O, Janssens S, McKenna W, Reichenspurn-
      er H, Trinquart L, Vilquin JT, Marolleau JP, Seymour B, Larghe-
      ro J, Lake S, Chatellier G, Solomon S, Desnos M, Hagège
      AA. 2008. The Myoblast Autologous Grafting in Ischemic
      Cardiomyopathy (MAGIC) trial: first randomized placebo-
      controlled study of myoblast transplantation. *Circulation*
      117:1189–200.

12.   Vacanti V, Kong E, Suzuki G, Sato K, Canty JM, Lee T. 2005.
      Phenotypic changes of adult porcine mesenchymal stem
      cells induced by prolonged passaging in culture. *J Cell
      Physiol* 2005:194–201.

13.   Foudah D, Redaelli S, Donzelli E, Bentivegna A, Miloso M,
      Dalprà L, Tredici G. 2009. Monitoring the genomic stability
      of in vitro cultured rat bone-marrow-derived mesenchymal
      stem cells. *Chromosome Res* 17:1025–39.

14. Momin EN, Vela G, Zaidi HA, Quiñones-Hinojosa A. 2010. The Oncogenic Potential of Mesenchymal Stem Cells in the Treatment of Cancer: Directions for Future Research. *Curr Immunol Rev* 6:137–48.

15. Itzhaki-Alfia A, Leor J, Raanani E, Sternik L, Spiegelstein D, Netser S, Holbova R, Pevsner-Fischer M, Lavee J, Barbash IM. 2009. Patient characteristics and cell source determine the number of isolated human cardiac progenitor cells. *Circulation* 120:2559–66.

16. Regulation (EC) No 1394/2007 of the European Parliament and of the Council of 13 November 2007 on advanced therapy medicinal products and amending Directive 2001/83/EC and Regulation (EC) No 726/2004.

17. Food and Drug Administration 21 CFR 1271 (2006).

18. Beltrami AP, Cesselli D, Bergamini N, Marcon P, Rigo S, Puppato E, D'Aurizio F, Verardo R, Piazza S, Pignatelli A, Poz A, Baccarani U, Damiani D, Fanin R, Mariuzzi L, Finato N, Masolini P, Burelli S, Belluzzi O, Schneider C, Beltrami AC. 2007. Multipotent cells can be generated in vitro from several adult human organs (heart, liver, and bone marrow). *Blood* 110:3438–46.

19. Nesselmann C, Ma N, Bieback K, Wagner W, Ho A, Konttinen YT, Zhang H, Hinescu ME, Steinhoff G. 2008. Mesenchymal stem cells and cardiac repair. *J Cell Mol Med* 12:1795–810.

20. Amado L, Saliaris A, Schuleri K, St. John M, Xie JS, Cattaneo S, Durand DJ, Fitton T, Kuang JQ, Stewart G, Lehrke S, Baumgartner WW, Martin BJ, Heldman AW, Hare JM. 2005. Cardiac repair with intramyocardial injection of allogenic mesenchymal stem cells after myocardial infarction. *Proc Natl Acad Sci USA* 102:11474–79.

21. Laugwitz KL, Moretti A, Caron L, Nakano A, Chien KR. 2008. Islet1 cardiovascular progenitors: A single source for heart lineages? *Development* 135:193–205.

22. Li L, Xie T. 2005. Stem cell niche: Structure and function. *Annu Rev Cell Dev Biol* 21:605–31.

23. Ayach BB, Yoshimitsu M, Dawood F, Sun M, Arab S, Chen M, Higuchi K, Siatskas C, Lee P, Lim H, Zhang J, Cukerman E, Stanford WL, Medin JA, Liu PP. 2006. Stem cell factor receptor induces progenitor and natural killer cell-mediated cardiac survival and repair after myocardial infarction. *Proc Natl Acad Sci USA* 103:2304–09.

24. Fazel S, CiminiM, Chen L, Li S, AngoulvantD, Fedak P, Verma S, Weisel RD, Keating A, Li RK. 2006. Cardioprotective c-kitþ cells are from the bone marrow and regulate the myocardial balance of angiogenic cytokines. *J Clin Invest* 116:1865–77.

25. Matsuura K, Nagai T, Nishigaki N, Oyama T, Nishi J, Wada H, Sano M, Toko H, Akazawa H, Sato T, Nakaya H, Kasanuki H, Komuro I. 2004. Adult cardiac Sca-1-positive cells differentiate into beating cardiomyocytes. *J Biol Chem* 279:11384–91.

26. Rosenblatt-Velin N, Lepore MG, Cartoni C, Beermann F, Pedrazzini T. 2005. FGF-2 controls the differentiation of resident cardiac precursors into functional cardiomyocytes. *J Clin Invest* 115:1724–33.

27. Goumans MJ, de Boer TP, Smits AM, van Laake LW, van Vliet P, Metz CH, Korfage TH, Kats KP, Hochstenbach R, Pasterkamp G, Verhaar MC, van der Heyden MA, de Kleijn D, Mummery CL, van Veen TA, Sluijter JP, Doevendans PA. 2008. TGF b1 induces efficient of human cardiomyocyte progeni-

tor cells into functional cardiomyocytes in-vitro. *Stem Cell Res* 1:138–49.

28. Engler A. J., Sen S, Sweeney HL, Discher DE. 2006. Matrix elasticity directs stem cell lineage specification. *Cell* 126:677–89.

29. Soliman S, Pagliari S, Rinaldi A, Forte G, Fiaccavento R, Pagliari F, Franzese O, Minieri M, Di Nardo P, Licoccia S, Traversa E. 2010. Multiscale three-dimensional scaffolds for soft tissue engineering via multimodal electrospinning. *Acta Biomater* 6:1227–37.

30. Murtuza B, Nichol JW, Khademhosseini A. 2009. Micro- and nanoscale control of the cardiac stem cell niche for tissue fabrication. *Tissue Eng Part B Rev* 15:443–54.

31. Even-Ram S, Artym V, Yamada KM. 2006. Matrix control of stem cell fate. *Cell* 126:645–47.

32. Di Felice V, Ardizzone NM, De Luca A, Marcianò V, Gammazza AM, Macaluso F, Manente L, Cappello F, De Luca A, Zummo G. 2009. OPLA scaffold, collagen I, and horse serum induce an higher degree of myogenic differentiation of adult rat cardiac stem cells. *J Cell Physiol* 221:729–39.

33 Marklein R. A., Burdick JA. 2010. Controlling stem cell fate with material design. *Adv Mater* 22:175–89.

34. Reilly GC, Engler A. J. 2010. Intrinsic extracellular matrix properties regulate stem cell differentiation. *J Biomech* 43:55–62.

35. Mandoli C, Pagliari F, Pagliari S, Forte G, Di Nardo P, Licoccia L, Traversa E. 2010. Stem cell aligned growth induced by CeO2 nanoparticles in PLGA scaffolds with improved bioactivity for regenerative medicine. *Adv Funct Mat* 20:1617–24.

36. Discher DE, Mooney DJ, Zandstra PW. 2009. Growth factors, matrices, and forces combine and control stem cells. *Science* 324:1673–77.

37. Vunjak-Novakovic G, Tandon N, Godier A, Maidhof R, Marsano A, Martens TP, Radisic M. 2010. Challenges in cardiac tissue engineering. *Tissue Eng Part B Rev* 16:169–87.

38. Forte G, Carotenuto F, Pagliari F, Pagliari S, Cossa P, Fiaccavento R, Ahluwalia A, Vozzi G, Vinci B, Serafino A, Rinaldi A, Traversa E, Carosella L, Minieri M, Di Nardo P. 2008. Criticatility of the biological and physical stimuli array inducing resident stem cell determination. *Stem Cells* 26:2093–103.

39. Jacot JG, McCulloch AD, Omens JH. 2008. Substrate stiffness affects the functional maturation of neonatal rat ventricular myocytes. *Biophys J* 95:3479–87.

40. Engler A. J., Carag-Krieger C, Johnson CP, Raab M, Tang HY, Speicher DW. Sanger JW, Sanger JM, Discher DE. 2008. Embryonic cardiomyocytes beat best on a matrix with heart-like elasticity: scar-like rigidity inhibits beating. *J Cell Sci* 121:3794–802.

41. Engelmayr GC Jr, Cheng M, Bettinger CJ, Borenstein JT, Langer R, Freed LE. 2008. Accordion-like honeycombs for tissue engineering of cardiac anisotropy. *Nat Mater* 7:1003–10.

42. Yeh YC, Lee WY, Yu CL, Hwang SM, Chung MF, Hsu LW, Chang Y, Lin WW, Tsai MS, Wei HJ, Sung HW. 2010. Cardiac repair with injectable cell sheet fragments of human amniotic fluid stem cells in an immune-suppressed rat model. *Biomaterials* 31:6444–53.

43. Aubin H, Nichol JW, Hutson CB, Bae H, Sieminski AL, Cropek DM, Akhyari P, Khademhosseini A. 2010. Directed 3D cell

alignment and elongation in microengineered hydrogels. *Biomaterials* 31:6941–51.

44. Zimmermann WH, Melnychenko I, Wasmeier G, Didié M, Naito H, Nixdorff U, Hess A, Budinsky L, Brune K, Michaelis B, Dhein S, Schwoerer A, Ehmke H, Eschenhagen T. 2006. Engineered heart tissue grafts improve systolic and diastolic function in infarcted rat hearts. *Nat Med* 12:452–58.

45. Martinez EC, Kofidis T. 2011. Adult stem cells for cardiac tissue engineering. *J Mol Cell Cardiol* 50:312–19.

46. Murry CE, Soonpaa MH, Reinecke H, Nakajima H, Nakajima HO, Rubart M, Pasumarthi KB, Virag JI, Bartelmez SH, Poppa V, Bradford G, Dowell JD, Williams DA, Field LJ. 2004. Haematopoietic stem cells do not transdifferentiate into cardiac myocytes in myocardial infarcts. *Nature* 428:664–68.

47. Oh H, Bradfute SB, Gallardo TD, Nakamura T, Gaussin V, Mishina Y, Pocius J, Michael LH, Behringer RR, Garry DJ, Entman ML, Schneider MD. 2003. Cardiac progenitor cells from adult myocardium: homing, differentiation, and fusion after infarction. *Proc Natl Acad Sci USA* 100:12313–18.

48. Smith RR, Barile L, Cho HC, Leppo MK, Hare JM, Messina E, Giacomello A, Abraham MR, Marbán E. 2007. Regenerative potential of cardiosphere-derived cells expanded from percutaneous endomyocardial biopsy specimens. *Circulation* 115:896–908.

49. Matsuura K, Honda A, Nagai T, Fukushima N, Iwanaga K, Tokunaga M, Shimizu T, Okano T, Kasanuki H, Hagiwara N, Komuro I. 2009. Transplantation of cardiac progenitor cells ameliorates cardiac dysfunction after myocardial infarction in mice. *J Clin Invest* 119:2204–17.

50. Okano T, Yamada N, Sakai H, Sakurai Y. 1993. A novel recovery system for cultured cells using plasma-treated polystyrene dishes grafted with poly (N-isopropylacrylamide). *J Biomed Mater Res* 27:1243–51.

51. Chen G, Hoffman AS. 1995. Graft copolymers that exhibit temperature-induced phase transitions over a wide range of pH. *Nature* 373:49–52.

52. Forte G, Pietronave S, Nardone G, Zamperone A, Magnani E, Pagliari S, Pagliari F, Giacinti C, Nicoletti C, Musarò A, Rinaldi M, Ribezzo M, Comoglio C, Traversa E, Okano T, Minieri M, Prat M, Di Nardo P. 2011. Human Cardiac Progenitor Cell Grafts as Unrestricted Source of Super-Numerary Cardiac Cells in Healthy Murine Hearts. *Stem Cells* (in press).

53. Dai W, Hale S. L., Kay G. L., Jyrala A. J., Kloner R. A. 2009. Delivering stem cells to the heart in a collagen matrix reduces relocation of cells to other organs as assessed by nanoparticles technology. *Regen Med* 4:387–95.

Chapter 12

Stem Cells: Present Perspectives and Future Challenges

Crystal M. Rocher, Sergey V. Bushnev* and Dinender K. Singla

Burnett School of Biomedical Sciences, College of Medicine, University of Central Florida, Orlando, FL 32817, *Neuroscience & Orthopaedic Clinical Research Institute, Florida Hospital, Orlando, FL 32804

For Correspondence:
Dinender K. Singla, PhD, FAHA
Associate Professor of Medicine
Burnett School of Biomedical Sciences
College of Medicine
University of Central Florida
Orlando, FL 32817

E-mail: dinender.singla@ucf.edu
Phone: 407-823-0953
Fax: 407-823-0956

1. Abstract

Stem cells are undifferentiated cells that remain in a self-renewal state and differentiate into a variety of body cell types when treated with appropriate growth factors. Their properties make them a beneficial resource in the area of regenerative medicine. Noticeably, carcinoma stem cells provided researchers the foundation necessary to establish the current field of stem cell research. A single cell from a teratocarcinoma tumor showed that it had the ability to differentiate into all the cell types found in these tumors. This discovery allowed for research to isolate and establish embryonic stem (ES) cells in the cell culture system. During development ES cells are responsible for deriving the three embryonic germ layers that make up every cell type found in the body. This ability to differentiate into a wide variety of cell types makes them great for therapeutic applications. Therefore, organs that once had limited regenerative potential, now have the ability to repair and reverse damage and injuries that may occur after adult stem cell treatment. Unfortunately, the benefits of these cells are dampened by the overwhelming controversy surrounding the moral and ethical concerns of destroying human embryos. In the last five years, an alternative to ES cells has been established that involves the reprogramming of somatic cells back to an embryonic-like state that are termed induced pluripotent stem (iPS) cells. Moreover, adult stem cells were initially thought to have limited plasticity, but current research suggests their potential to transdifferentitate, or differentiate into cell types of different lineages. In this chapter, we will discuss ES, iPS, and adult stem cell properties, applications, and associated challenges.

2. Introduction

The theory of tissue regeneration and repair has always remained interesting and puzzling throughout the existence of humans and animals [4,7]. This puzzle became more complex when it was learned that some fish and amphibians have the potential to repair or partially repair damaged tissue or organs [7]. Zebrafish, for example, have the ability to regenerate their fins, spinal cord, and retina [33]. It was more recently shown that they can also regenerate their own heart after injury [33]. Similarly, MRL mice, when given the standard ear-hole punch used to identify age and sex in animal laboratories, were able to fully heal the wound and regenerate new skin and cartilage within four weeks [1,8]. Subsequently, when myocardial injury was generated in the MRL mice, autorecovery occurred within a few weeks, suggesting MRL mice have regenerative capability. In addition, Porrello et al. recently showed that the neonatal mouse heart has the ability to regenerate up to a week after birth [32]. Moreover, human and most large animal adult hearts, do not have regenerative potential after injury; therefore, alternate strategies are needed to repair the injured myocardium [7].

In this regard, stem cells have been identified as potential candidates, as these cell types can self-renew in the body and stay in an undifferentiated state in the cell culture [46]. There are three major categories of stem cells: embryonic stem (ES) cells, induced pluripotent stem (iPS) cells, and adult stem cells (ASCs). Traditionally, the treatments for various medical conditions or diseases included antibiotics or pharmaceutical interventions, which do not always successfully solve the problem [20]. Therefore, these interventions are not considered very effective as mortality and morbidity is still on the rise [20]. Stem cells, which have the potential to repair and regenerate the organ, can be a more effective way to treat disease. In such instances, stem cells, whether they are ES, iPS, or adult, would have the ability to be transplanted into the specific area, incorporate itself, and begin differentiating into the desired cell types, which would allow for regeneration [7,36–38]. Organs such as the brain and heart, which traditionally were thought not to have the ability to regenerate, would now be able to regenerate following transplantation of stem cells [7]. This chapter will delve into the history of stem cells, the therapeutic applications of stem cells, and ways to circumvent the challenges and issues that plague this cell therapy.

3. Embryonic Stem Cells

3.1 History and background

Carcinoma stem cell research first began in the 1950s on teratomas and teratocarcinomas, which are benign and malignant tumors found in the gonads [40]. These tumors consist of adult tissue and organs and usually are described as containing hair, skin, muscles, and bone [40]. Two researchers were able to see how a single cell from a tumor could go on to form all the cell types found in these tumors after injection back into the animal [40]. The obtained data from these animals was significant because it was the first findings that described the nature of a simple cell as a pluripotent stem cell [40]. However, the mechanism of self-renewal and thereby differentiation into specific cell types in tumors was completely unknown [40]. In addition, the formation of structures that resembled early embryos called embryoid bodies (EBs) was observed, and it was anticipated that these tumors would be identical to mouse embryos in origin and development [40]. Furthermore, this study gave rise to embryonal carcinoma (EC) cell research as EC cells were derived and isolated from implantation of inbred mouse embryos into an extra-uterine site of a compatible host [21,40]. ES cells were seen as a potential useful alternative to mimic mouse embryos in order to study embryonic cell differentiation in the cell culture system based on their similarities to ES cells in vivo [21]. Additionally, these cells are pluripotent, which means they are able to differentiate into all three germ layers present during development [21,22].

In 1981, Evans and Kaufman isolated ES cells from a mouse blastocyst [10]. In this study they mated 129 SvE mice strains for the isolation of ES cells [10]. After confirming mating plug had occurred in these animals, they delayed the implantation of the blastocyst to the uterine wall for four to six days by performing an ovariectomy between the second and third day of pregnancy in order to increase growth of ICM so that adequate numbers of ES cells could be isolated [10]. This was an experimental way of delaying the implantation by altering the maternal hormonal conditions [10]. After that, the blastocysts were recovered and placed on tissue culture plates and allowed to differentiate into trophoblast cells [10]. The EC-like cells were derived from the inner cell mass (ICM) cells and manually picked and then placed on gelatin coated plates containing a STO fibroblast feeder layer [10]. Interestingly, they noticed that the cells derived from the ICM, with unique characteristics, resembled EC cells. Furthermore, generation of tumor formation in vivo and embryoid body formation in vitro was required to further confirm these cells' pluripotency in nature, which is the major characteristic of ES cells [10]. After confirmation of these characteristics, they announced the isolation of ES cells.

Next, in 1981, another group expanded on the work of Evans and Kaufman. In this study, superovulated female mice were mated and 76 hours after mating plug was observed, blastocysts were flushed from the uterus [21]. The blastocysts were maintained in DME medium in order to allow them to fully expand, and an immunosurgery protocol, a similar procedure used by Solter et al. to isolate mouse ES cells, was used to extract the ICM [21,41]. The ICM cells were grown on tissue culture plates containing a feeder layer of STO fibroblasts, and ES cell colonies were formed with the help of EC cell conditioned media (CM) [21]. This study theorized that certain factors in the CM would aid the ICM cells into ES cell-like colonies [21]. Again, like in Evans and Kaufman's study, this new study was able to prove these cells were indeed pluripotent stem cells by their morphology and based on their formation of tumors in vivo and EBs in vitro [21].

Now that a method was established on how to isolate ES cells from a mouse embryo, a method to maintain these newly isolated cells in the cell culture system for a long period of time was another major challenge that needed to be determined. The method that was used required the addition of a fibroblast feeder layer followed by EC cell CM [21]. Studies had shown that factors located in the CM of these cells allowed the ES cells to stay in an undifferentiated state, but these factors still remained unknown [21]. Furthermore, there were challenges with this method such as safety issues involving the use of EC cells. Leukemia inhibitory factor (LIF) was discovered in 1988 in order to maintain ES cells in an undifferentiated state in vitro for long periods of time [50]. Williams et al. identified increased concentrations of LIF was found in the CM of fibroblast feeder layers and in EC cells [50]. Moreover, they also showed the presence of a LIF receptor on ES cells [50]. In addition they found that when these same ES cells were maintained without the presence of LIF, they differentiated into various body cell types within three to five days [50]. Therefore, these studies suggested that LIF is a key compound in the CM required for ES cell self-renewal and maintenance in the cell culture system [50]. With this new evidence, mouse ES cells could be maintained in vitro in the presence of LIF without STO cells or EC cell CM [50].

Next, Thompson et al. isolated nonhuman primate ES cells, which were the closest to mimic human EC cells [47]. Their group examined mouse ES and EC cells positively expressing cell surface markers and alkaline phosphatase stainings [47]. They confirmed the presence of SSEA-1, but not SSEA-3 or 4, Tra-1-60, or Tra-1-81 in mouse ES and EC cells [47]. Moreover, primate ES cells and human EC cells positively expressed SSEA-3 or 4, Tra-1-60, and Tra-1-81, but not SSEA-1 [47]. In addition, primate ES cells were shown to be able to maintain a self-renewal state, had a normal karyotype, and had the ability to differentiate into the three embryonic germ layers [47].

After isolating primate ES cells, Thompson et al. were also able to isolate human ES cells from donated embryos produced through in vitro fertilization [46]. The embryos were grown until the blastocyst stage and then the ES cells were isolated using the immunosurgery protocol [46]. They observed that the human ES cells had the same morphology as primate ES cells; the karyotype was normal, and the cells were able to proliferate and stay undifferentiated [46]. Moreover, human ES cells have higher levels of telomerase activity as compared to somatic cells, which affects the life-span of a cell [46]. These cell also have positively expressed SSEA-3 and alkaline phosphatase but not SSEA-1 as observed in primate ES cells and human EC cells [46]. Human ES cells were able to successfully differentiate into all three embryonic germ layers like mouse ES cells and primate ES cells, but like primate ES cells, were not able to stay undifferentiated in the presence of LIF [46]. Human ES cells were found to stay in an undifferentiated state when grown on a mouse embryonic fibroblast (MEF) feeder layer with the help of basic fibroblast growth factor (bFGF) or feeder free when the tissue culture plate was coated with either matrigel or laminin and MEF conditioned media [5]. Further research showed that high concentrations of bFGF yielded the removal of the MEF feeder layer [19]. Currently, a new media called TeSR1 is on the market for human ES cells that does not include any components of animal origin and allows for the cells to maintain in a self-renewal state without the help of a feeder layer [19].

3.2 Embryonic germ layer cell types

During fertilization, the germ cells undergo meiosis to give rise to the male and female gametes, which fuse together to form the zygote [7]. The zygote is totipotent, meaning it has the ability to derive an entire organism [7,46]. The zygote will then cleave or divide, also known as mitosis, until it becomes the blastocyst [7]. Located in the blastocyst is the ICM, from which ES cells are derived [7]. The embryo has now gone from being totipotent to pluripotent with the differentiation of the ES cells into the three embryonic germ layers found in the gastrula as shown in Figure 1 [7,48]. The three germ layers are comprised of the endoderm, mesoderm, and ectoderm and play an important role in the development of the embryo [46]. The endoderm, or internal layer, is responsible for the development of cell types that generate the digestive tract and lung, such as lung cells, thyroid cells, and pancreatic cells [25,49]. The mesoderm, or middle layer, is responsible for cells types that generate the connective tissue, blood, bone, muscle, and fat, such as adipocytes, osteoblasts, and chondrocytes [25,43,49,49]. The ectoderm, or external layer, is responsible for cells types that generate the skin and brain, such as keratinocytes, melanocytes, and neurons [15,25,49].

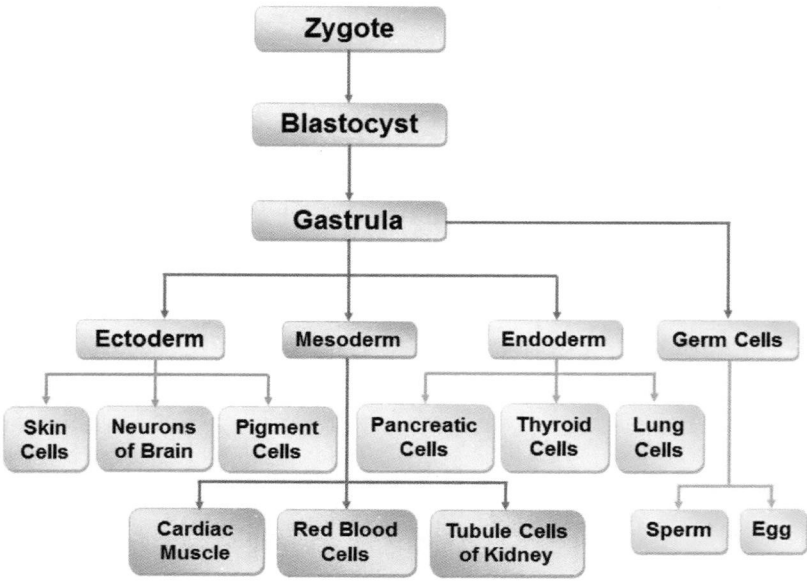

Figure 1. Flow chart depicts the development of an embryo from zygote to blastocyst to gastrula and finally to the three germ layers.

3.3 Primordial germ cells

During differentiation into the germ layers, cells derived from the ICM also generate primordial germ cells (PGCs), also shown in Figure 1 [48]. PGCs are the precursor cells that give rise to germ cells which in turn develop into the gametes that are responsible for fertilization [7]. They are located in the genital ridges of the developing embryos [11]. PGCs migrate to the gonads and determine the sexual orientation of the developing embryo [29]. Embryonic germ (EG) cells are derived from these cells and exhibit pluripotency and many other properties similar to ES cells [35]. Because they are pluripotent, they have the ability to generate all three embryonic germ layers and can form EBs to initiate differentiation [35]. Furthermore, they exhibit a stable karyotype, the ability to remain in a self-renewed state, and high expression of alkaline phosphatase [35].

3.4 Induced pluripotent stem cells

The possibilities surrounding ES cells for clinical applications are very optimistic, but certain challenges such as ethical issues as well as teratoma formation following transplantation in the organ limit this enthusiasm [44]. An alternative needed to be discovered to generate cells which could overcome these limitations. In 2006, Takahashi et al. were able to successfully generate embryonic-like cells from MEF and adult mouse tail-tip fibroblasts [44,45]. In this study they tested twenty-four genes and discovered four transcription factors (Oct3/4, Sox2, c-Myc, and Klf4) that were essential to maintain pluripotency once they transfected the somatic cells using the retrovirus method [45]. They named these cells induced pluripotent stem (iPS) cells [45]. iPS cells are somatic cells that are genetically reprogrammed to an ES cell-like state [23,44,45,51]. iPS cells, like ES cells, have the capability to differentiate into all three of the embryonic germ layers (ectoderm, endoderm, and mesoderm) as well as remain in a self-renewal state, as shown in Figure 2 [44,45,51]. They observed that the mouse iPS cells were exactly similar to mouse ES cells in morphology, gene expression, teratoma formation, and proliferation and were able to successfully develop mouse chimeras when the iPS cells were transplanted into the blastocyst of a host female [44]. This data triggered more research to generate human iPS cells, which, like mouse iPS cells, would have all the same properties as human ES cells. In 2007, Takahashi et al. successfully accomplished this. For the human iPS cells, they used adult human dermal fibroblasts as the somatic cells to be transfected and, like the mouse iPS cells, used the same four transcription factors of Oct3/4, Sox2, c-Myc, and Klf4

[44]. This demonstrated that although the factors involving pluripotency were the same between mouse and human iPS cells, the factors needed to sustain the cultures in an undifferentiated state had not changed [44]. These human iPS cells were also able to exhibit the same morphology, gene expression, teratoma formation, proliferation, and telomerase activity as human ES cells [44]. The generation of iPS cells proves to be an optimistic alternative and tool for advances in regenerative medicine.

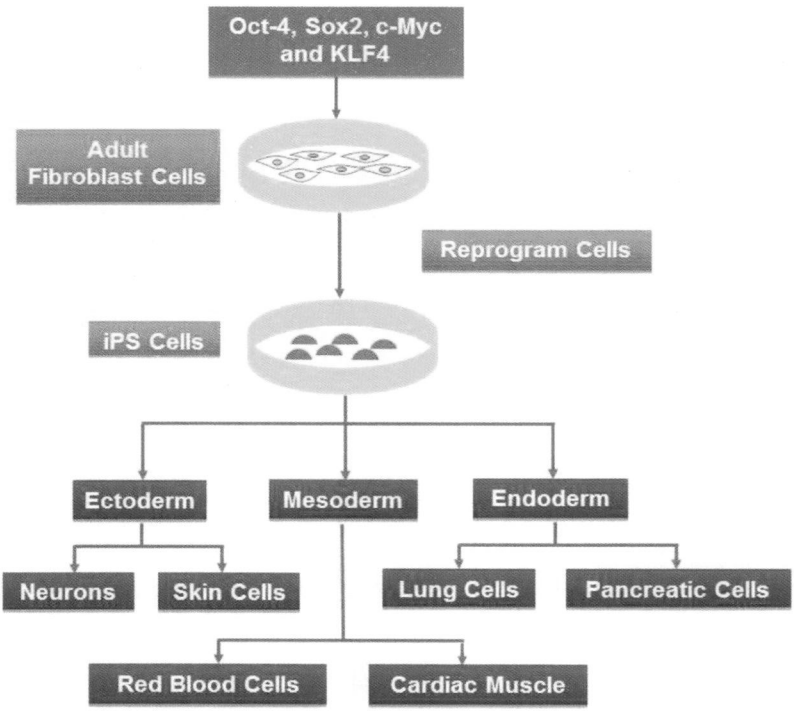

Figure 2. Diagram illustrating the transfection of somatic cells into iPS cells. Adult fibroblasts are transfected with four transcription factors (Oct-4, Sox2, c-Myc, and Klf4) and reprogrammed into induced pluripotent stem cells, which can differentiate into cell types found in all three embryonic germ layers.

4. Adult Stem Cells

4.1 History and background

Adult stem cell (ASC) research has been going on for decades, and these cells have been successfully used as a therapeutic application. One of the first areas to show potential was the bone marrow, which contains two types of stem cells [3]. In 1909, a Russian military doctor, Alexander Maximow, discovered the first stem cell in bone marrow, which would come to be termed a hematopoietic stem cell (HSC) [12]. This cell was observed to resemble a lymphocyte with the ability to migrate throughout the blood and either stay in an undifferentiated state or differentiate into the necessary blood cell types, such as T-cells and monocytes [12]. By the 1940s, HSC transplantation following cancer treatments was becoming more prominent [18]. Data obtained from these patients suggested that their disease conditions had significantly improved, which had a major impact on therapeutic applications and regenerative medicine. In 1968, Friedenstein and his colleagues discovered the existence of another stem cell located in the bone marrow [2]. These cells had the ability to derive the cartilage, bone, fat, and other connective tissue generated from the bone marrow [2]. These cells, compared to HSCs, had a more fibroblast-like morphology showing their origin to the stroma of the bone marrow [4]. Eventually these stem cells would be termed mesenchymal stem cells (MSCs) [4]. Additionally, neural stem cells (NSCs), another ASC that showed potential, were observed in the 1960s with the discovery of neurogenesis in the hippocampus of rats and guinea pigs [9]. By the 1990s, stem cells discovered in the central nervous system where shown to be able to remain in a self-renewal state for an unlimited amount of time but also differentiating into three specific cell types: neurons, astrocytes, and oligodendrocytes [15]. These cell types were eventually termed NSCs [15]. All of these advances in stem cell research showed that adult stem cells could be isolated from various tissues in the body.

4.2 Adult stem cell plasticity

ASCs are found in most of the tissues or organs of the mammalian species and can either stay in an undifferentiated state or differentiate into the cell types of the tissues and organs of origin. There are many types of ASCs, as shown in Figure 3. Many believed that ASCs had no plasticity compared to ES and iPS cells, but recent studies have shown that to not be completely true [49]. Some ASCs under certain conditions may be able to have a wider effect on cell differentiation [49]. Transdifferentiation is the transformation of a cell of one lineage to a cell of a completely different lineage [49]. For example, NSCs have been shown to differentiate into blood cells under specific conditions, and HSCs have differentiated into glial cells [9]. ASCs have the ability to lose all of their original properties and functions and gain those of the new cell type [49]. These cells are proving to have more plasticity then once believed [9].

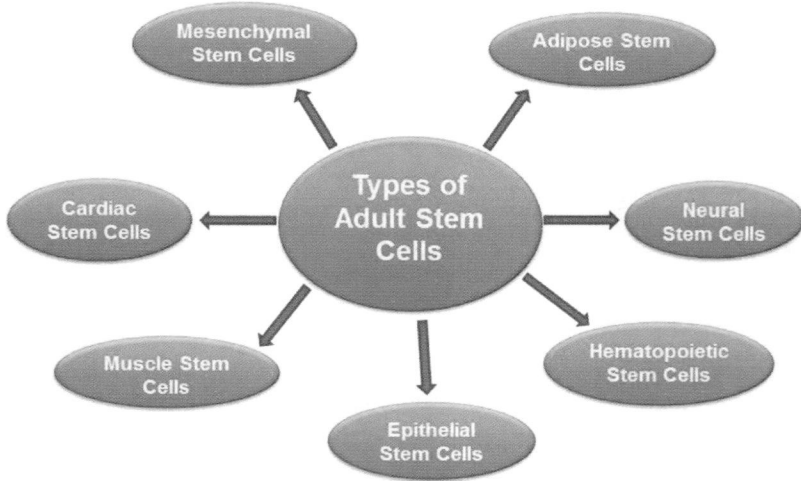

Figure 3. Illustrates different types of adult stem cells found in the human body.

4.3 Types of adult stem cells and their plasticity

MSCs are located in the bone marrow and are responsible for deriving the tissue of the mesenchyme such as bone, cartilage, muscle, and adipose [31]. Being a stem cell, they have the ability to either remain in a self-renewal state or differentiate into the multiple cell lineages. Furthermore, MSCs have proven to have the ability to transdifferentiate [14]. For example, MSCs can differentiate into brain astrocytes [14].

Adipose stem cells, also called fat stem cells, are derived from the mesenchyme and are seen as an optimistic alternative to MSCs, which experience many challenges regarding its clinical use [52]. Adipose stem cells are also responsible for deriving fat, bone, cartilage, and muscle [43]. Contrary to most other adult stem cells, they are more readily available through the surgical technique known as liposuction [43]. Moreover, these cells have been shown to successfully generate iPS cells [43].

HSCs are located in the bone marrow and are responsible for deriving all the blood cells [12]. There are two main types of blood cells which consist of lymphoid and myeloid progenitor cells [12,25]. The lymphoid progenitor cells generate the natural killer cells, T-cells, and B-cells, while the myeloid progenitor cells derive the monocytes, erythrocytes, megakaryocytes, and granulocytes (Figure 4) [12]. These stem cells also have the ability to transdifferentiate under specific conditions [9]. For instance, they have been proven to differentiate into cardiomyocytes and hepatocytes [14].

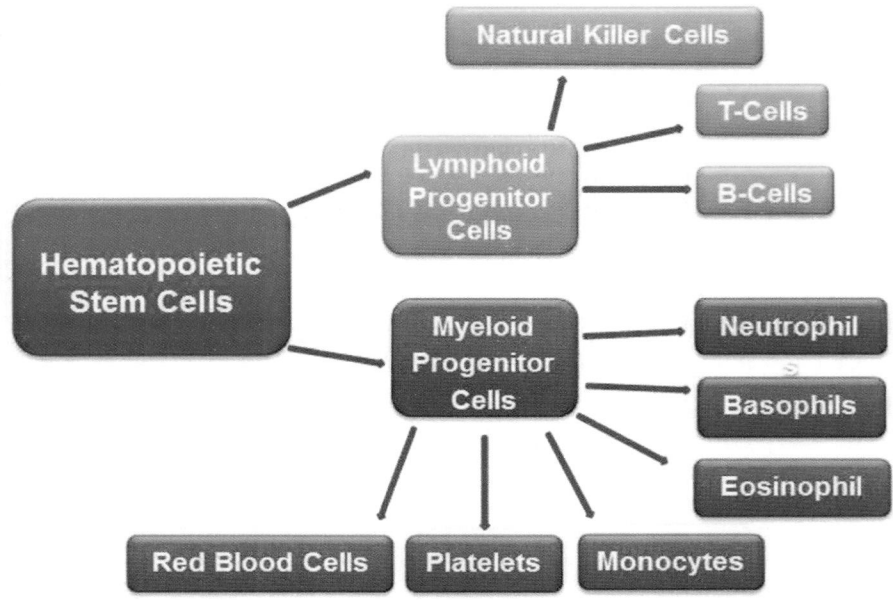

Figure 4. Depicts the cell types that are derived from hematopoietic stem cells. Two progenitor cell types are derived from hematopoietic stem cells, which go on to differentiate into the various blood cell lineages.

NSCs are located in the brain and central nervous system and differentiate into neurons, astrocytes, and oligodendrocytes [15]. Neural crest stem cells are responsible for the development of the peripheral nervous system, the craniofacial mesenchyme, and melanocytes [15]. Neural crest stem cells differ from neural stem cells in their ability to travel to different areas and their role in mesenchymal lineages [15]. Moreover, NSCs, like HSCs, are shown to have the ability to transdifferentiate under appropriate conditions [9].

Epithelial stem cells are located in the epithelia which comprise many areas of the human body [6]. The epithelium consists of sheets of tightly linked cells that cover the surfaces and lining of the body, such as the epidermis and the corneal epithelium, as well as the digestive and respiratory epithelia [6]. The epithelia act as protection from outside factors and regulate what gets in and out, such as water and nutrient absorption and granular secretions [6].

Cardiac stem cells (CSCs) are found in the heart and differentiate into cardiomyocytes, smooth muscle, and endothelial cells [26]. In vivo experiments have shown that when injected into the outside border of the infarct region of a myocardial infarction, the CSCs were able to repair the injured myocardium by about 25% [26]. CSCs give hope that the heart has self-renewal potential and can be used as a therapeutic application.

Muscle stem cells, or satellite cells, are precursor cells located in skeletal muscle [14]. When the muscle is injured, these stem cells become myoblasts, which eventually differentiate into myotubes and then become muscle fibers [14].

4.4 Differences from ES cells

There are many differences between ASCs and ES cells. As stated above ASCs do not have plasticity as ES cells do, which are pluripotent so they can theoretically differentiate into all the cell types in the body [7]. ASCs are also harder to isolate and culture in vitro [7]. They also have less telomerase activity, which is probably related to the challenges in culturing since telomerase is directly related to life-span [7,46]. The advantages of ASCs, though, are that there are not any ethical concerns surrounding these cells. Both adult and ES cells have the risk of teratoma formation following transplantion [7]. Furthermore, a negative immune response would be less likely in ASCs because they are autologous cells for transplantation, whereas ES cells are not autologous cells [14].

5. Therapeutic Applications

5.1 ES cells

ES cells could become a potential tool to fight and prevent diseases, injuries that the human body sustains, and genetic abnormalities. For instance, ES cells have shown the ability to differentiate into all the cell types found in the heart and the ability to engraft to the injured myocardium [36]. If a patient suffers from a myocardial infarction, the transplantation of ES cells or of cardiomyocytes derived from ES cells could result in repair of the injured myocardium [11,36]. Furthermore, several factors have been identified that enhance the differentiation of ES cells to cardiac cell types such as transforming growth factor-β2 (TGF-β2) [39]. Mouse ES cells treated with TGF-β2 were shown to have a percentage of beating EBs of 38–50%, while the ES cells that were not treated only had a percentage of 2–15% [39]. ES cells also have the ability to affect the treatment of diabetes. Their ability to differentiate into pancreatic B-cells or insulin-containing cells is beneficial for the treatment of type I diabetes [25,42]. Furthermore, studies have shown that insulin-containing cells derived from ES cell differentiation are able to normalize blood glucose in diabetic mice [42]. Studies have also been successful with deriving pancreatic B-cells from human ES cells [1].

The brain, like the heart, was initially thought to not have much regenerative potential based on the limited amount of progenitor cells located in these areas, but transplantation of ES cells has the ability to change this. As stated above, their ability to differentiate into many cell types makes them a beneficial therapeutic application for neural degenerative diseases and spinal cord injuries as well. Reubinoff et al. was able to show that ES cells can differentiate into neural progenitor cells, which go on to differentiate into neurons, astrocytes, and oligodendrocytes [34]. Spinal cord injuries could be treated by introducing new neurons derived from ES cells that would regenerate the area of injury and enhance remyelination [25]. Mouse and rat spinal cord injury models were shown to have recovery of the hindlimb following transplantation of neural progenitor cells derived from ES cells [16,23,25]. Furthermore, human ES cells have been shown to differentiate into dopamine neurons, which can potentially be very beneficial for the treatment of diseases like Parkinson's Disease [30]. Perrier et al. was able to achieve this by the co-culture of human ES cells on a MS5 stroma along with treatment of fibroblast growth factor 8 (FGF8) and Sonic hedgehog (SHH) [30]. This area is still very new and requires more research, but some have shown how specific transcription and growth factors can direct the cells to a certain direction [11].

ES cells have the potential to greatly improve modern medicine and can be used as a tool to understand the human body and understand the mechanisms involving diseases and genetic disorders, but more research must be done to fully understand the complexity of these cells and of their role in development.

5.2 iPS cells

Although ES cells have the ability to differentiate into all three primary germ layers, ethical concerns and the fear of teratoma formation make them difficult for use in a clinical

347

setting. iPS cells, on the other hand, have all the same properties as ES cells without the controversy. They are reprogrammed to an ES cell–like state from somatic cells and have the ability to enhance the opportunities of regenerative medicine [27]. iPS cells would be able to circumvent the challenges associated with ES cells by removing the fear of an immunoresponse based on patient-specific cell lines, eliminate stem cell shortages, and reprogram tissue to create progenitor cells [27]. Furthermore, iPS cells have the ability to repair genetic abnormalities from sequence defects by homologous recombination [27].

Earlier, the positive effects of adult stem cells were shown for the treatment of sickle cell anemia. iPS cells have the potential to further improve on what hematopoietic stem cells can do. For example, Hanna J et al. used a mouse model for sickle cell anemia [13]. Tissue from that model was used to be reprogrammed to generate iPS cells, which were then corrected of the sickle cell gene [13,27]. These cells were then introduced along with hematopoietic cells cultured in vitro, and pathognomonic effects were reversed [13,27]. The effects of iPS cells on the brain were also shown by an animal model that, for Parkinson's Disease, showed the effects of iPS cells on producing dopamine cells [27]. Neurons derived from iPS cells were successfully transplanted and showed neuronal activity along with the production of dopamine [27]. This model successfully showed the improvement of symptoms associated with Parkinson's disease along with no signs of teratoma formation [27]. Furthermore, the effects of iPS cells can be seen for ischemic heart disease. Transplanted iPS cells showed repaired function and improved electrical stability, and regeneration of injured tissue was observed up to four weeks post-injury [27]. Moreover, Singla et al. showed that iPS cells can differentiate into cardiomyocytes, regenerate the infarcted myocardium following a myocardial infarction (MI), and inhibit apoptosis and fibrosis up to two weeks post MI [38]. iPS cells are an ideal alternative to ES cells and have the potential to enhance regenerative medicine.

5.3 Adult stem cells

ASCs provide another alternative to ES cells and are greatly important to the field of regenerative medicine themselves. As stated above many have the ability to either be reprogrammed into iPS cells or transdifferentiate under appropriate conditions [14,45,49]. Many researchers have shown how stem cells from the bone marrow, for instance, can differentiate into cell types necessary for the brain or heart [14]. This property could be beneficial for ASCS that are difficult to isolate and culture on a large scale for clinical purposes [7]. Moreover, it gives insight into genetic modification of stem cells [49]. Their natural ability to function as a repair system for injured tissue or organs indicates the potential they can have in a clinical setting. Furthermore, unlike ES cells, ASCs do not suffer from the same challenges such as ethical concerns, immunorejection, or teratoma formation [14]. All of these properties make them a beneficial resource for therapeutic applications.

There are many diseases and conditions in which ASCs could be used and are used as a treatment. As stated before, HSCs have the responsibility of deriving blood cells, which makes them a great resource for treating patients with immunodeficient or autoimmune diseases [24]. By introducing new T and B-cells that are vital to the immune system, patients not only are able to treat and repair what is affecting them but also end up with a better immune system [24]. It can also be used for treatment of specific cancers, such as leukemia, or sickle cell anemia [12]. In the latter, sickle cell anemia results from red blood cells getting a sickle shape, which causes them to be weak and only survive twenty days as opposed to the normal life span of one hundred twenty days [27]. Thus, the bone marrow is not able to produce as many red blood cells as necessary and results in not enough oxygen getting to the rest of the body, since hemoglobin attached to red blood cells carry this oxygen, and eventually leads to hypoxia [27]. HSCs, with their ability to differentiate into blood cells, could be used to reverse this problem and allow the bone marrow to produce more red blood cells [27]. NSCs have been shown to differentiate into the cell types found in the brain and central nervous system in both rodent and human brains, making them a useful tool for the treatment of neurodegenerative diseases [9,15,24]. For example, when NSCs are transplanted into the brain, they have been shown to repair damage to the brain and spinal cord and restore function and have most importantly had the ability to migrate to the damaged area [24]. This allows the NSCs to not only repair the damage but introduce neuroprotective factors [24]. In cases of heart disease, CSCs, as stated before, are essential for the recovery of damaged heart tissue [26]. In recent animal studies, it has been demonstrated that transplanted CSCs in the infarcted heart can regenerate with newly differentiated heart cell types as well as improved cardiac function [26].

6. Challenges and Ethical Concerns

There are many challenges and concerns that still plague stem cells. Many countries in the world realize that it is deplorable to use human embryos, even if these embryos are donated or were meant to be discarded. Many religious groups do not condone abortion or even in vitro fertilization and feel this research will eventually lead to the cloning of human beings. Furthermore, many believe life begins at conception and feel that by destroying a human embryo, you are destroying life and as such should not be federally funded. Conversely, people who suffer from ailments and diseases that could possibly be cured by stem cells, or at least lead to an improved way of life, feel they should not be punished based on the moral beliefs of a portion of the population. This is a contested issue that has been affecting those on both sides of the issue for decades.

Moral concerns are not the only issues that affect stem cell research; many ethical and safety issues affect this research as well. For example, can a researcher guarantee that, when transplanted, stem cells will not begin to proliferate and differentiate uncontrollably or lead to the formation of teratomas. This fear is part of the reason why ES cells or carcinoma cells are not used for clinical purposes and why some fear the effects of media consisting of components of animal origin [19]. Furthermore, there is a concern as to what side effects may come from directly transplanting the stem cells or possible negative immune responses that could occur.

Because of these concerns and proven challenges, tests must be done to guarantee cells used for clinical purposes are one hundred percent safe [7]. Furthermore, any stem cells used must be tested on animal models to ensure there will be no formation of teratomas and that they will maintain normal physiological function [7]. When it comes to differentiation, researchers must be positive that these cells will differentiate into the correct cell type in regards to the specific application required to generate a particular organ [7]. When it comes to adult stem cells, methods must be created to ensure large-scale production for therapeutic applications as well as no genetic abnormalities occur when grown in the cell culture system. Recent studies suggest bone marrow stem cells, when passaged at low rates, contain genetic abnormalities and induce teratomas in the hearts of animals and in diabetic animals. Moreover, fear of immunorejection in relation to ES cells must be overcome to ensure stable transplantation and successful regeneration of tissue [7].

7. Ways to Circumvent These Challenges

The challenges associated with ES cells may never be resolved without the help of alternative methods. People may always have an opinion on the morality of using human embryos, thereby giving a negative connotation to ES cell research. Although there are many benefits to ES cell research, there are other more difficult challenges that must be overcome besides ethical ones. For instance, as stated above, the two most prevalent involve immunorejection of ES cells following transplantation and the possible formation of teratomas. iPS cells present a possible solution to these problems. The properties and benefits of these cells were discussed in previous sections and are shown to have great potential to be a replacement. Not only can they differentiate into all three primary germ layers, like ES cells, but they can be genetically modified and can be patient-specific [27]. Genetic modification can allow for positive directed differentiation of cells as well as program the cells to be used as a delivery system, as discussed above [24].

When it comes to teratoma formation, the use of adult stem cells could be an alternative. The advancement in the cultivation of these stem cells and the fact that they can transdifferentiate shows they have great potential in a clinical setting. Furthermore, research has been done to see if the number of stem cells transplanted has an effect on teratoma formation. Lee et al. showed just that. When mice received an intramyocardial injection containing 1×10^6 ES cells, teratomas were formed after three to four weeks post-injection [17]. Furthermore, teratoma formation was not just observed in the heart, ES cells were observed to have traveled to other organs such as the kidney and liver [17]. Although the formation of teratomas is not beneficial, this was further proof that stem cells have the ability to migrate naturally. Next they looked at the effects of injecting 1×10^5 human ES cells. Only two out of seven showed teratoma formation, but after four weeks the rate of teratoma growth significantly increased [17]. Conversely, injections of 1×10^4 or less showed no formation of teratomas [17]. This would prove beneficial for the use of ES and iPS cells. Moreover, Nussbaum et al. showed that when differentiated cells are purified from ES cells, they are less likely to form teratomas, which shows the importance of finding methods to better enhance cell selection [28].

References

1. Bedelbaeva K., Snyder A., Gourevitch D., Clark L., Zhang X. M., Leferovich J., Cheverud J. M., Lieberman P., and Heber-Katz E. 2010. Lack of p21 expression links cell cycle control and appendage regeneration in mice. *Proc Natl Acad Sci U S A* 107: 5845-5850.

2. Bianco P. and Gehron R. P. 2000. Marrow stromal stem cells. *J Clin Invest* 105: 1663-1668.

3. Bianco P., Riminucci M., Gronthos S., and Robey P.G. 2001. Bone marrow stromal stem cells: nature, biology, and potential applications. *Stem Cells* 19: 180-192.

4. Bianco P., Robey P. G., and Simmons P. J. 2008. Mesenchymal stem cells: revisiting history, concepts, and assays. *Cell Stem Cell* 2: 313-319.

5. Bishop A. E., Buttery L. D., and Polak J. M. 2002. Embryonic stem cells. *J Pathol* 197: 424-429.

6. Blanpain C., Horsley V., and Fuchs E. 2007. Epithelial stem cells: turning over new leaves. *Cell* 128: 445-458.

7. Bongso A. and Richards M. 2004. History and perspective of stem cell research. *Best Pract Res Clin Obstet Gynaecol* 18: 827-842.

8. Clark L. D., Clark R. K., and Heber-Katz E. 1998. A new murine model for mammalian wound repair and regeneration. *Clin Immunol Immunopathol* 88: 35-45.

9. Eridani S. 2002. Stem cells for all seasons? Experimental and clinical issues. *J R Soc Med* 95: 5-8.

10. Evans M. J. and Kaufman M. H. 1981. Establishment in culture of pluripotential cells from mouse embryos. *Nature* 292: 154-156.

11. Gepstein L. 2002. Derivation and potential applications of human embryonic stem cells. *Circ Res* 91:866–876.

12. Gunsilius E., Gastl G., and Petzer A. L. 2001. Hematopoietic stem cells. *Biomed Pharmacother* 55: 186-194.

13. Hanna J., Wernig M., Markoulaki S., Sun C. W., Meissner A., Cassady J. P., Beard C., Brambrink T., Wu L. C., Townes T. M., and Jaenisch R. 2007. Treatment of sickle cell anemia mouse model with iPS cells generated from autologous skin. *Science* 318:1920–1923.

14. Hombach-Klonisch S., Panigrahi S., Rashedi I., Seifert A., Alberti E., Pocar P., Kurpisz M., Schulze-Osthoff K., Mackiewicz A., and Los M. 2008. Adult stem cells and their trans-differentiation potential--perspectives and therapeutic applications. *J Mol Med (Berl)* 86: 1301-1314.

15. Kennea N. L. and Mehmet H. 2002. Neural stem cells. *J Pathol* 197: 536-550.

16. Kimura H., Yoshikawa M., Matsuda R., Toriumi H., Nishimura F., Hirabayashi H., Nakase H., Kawaguchi S., Ishizaka S., and Sakaki T. 2005. Transplantation of embryonic stem cell-derived neural stem cells for spinal cord injury in adult mice. *Neurol Res* 27: 812-819.

17. Lee A. S., Tang C., Cao F., Xie X., van der B. K., Hwang A., Connolly A. J., Robbins R. C., and Wu J. C. 2009. Effects of cell number on teratoma formation by human embryonic stem cells. *Cell Cycle* 8: 2608-2612.

18. Little M. T. and Storb R. 2002. History of haematopoietic stem-cell transplantation. *Nat Rev Cancer* 2: 231-238.

19. Ludwig T. E., Levenstein M. E., Jones J. M., Berggren W. T., Mitchen E. R., Frane J. L., Crandall L. J., Daigh C. A., Conard K. R., Piekarczyk M. S., Llanas R. A., and Thomson J. A. 2006. Derivation of human embryonic stem cells in defined conditions. *Nat Biotechnol* 24: 185-187.

20. Mancini D. and Lietz K. 2010. Selection of cardiac transplantation candidates in 2010. *Circulation* 122: 173-183.

21. Martin G. R. 1981. Isolation of a pluripotent cell line from early mouse embryos cultured in medium conditioned by teratocarcinoma stem cells. *Proc Natl Acad Sci U S A* 78: 7634-7638.

22. Martin G. R. and Evans M. J. 1975. Differentiation of clonal lines of teratocarcinoma cells: formation of embryoid bodies in vitro. *Proc Natl Acad Sci U S A* 72: 1441-1445.

23. McDonald J. W., Liu X. Z., Qu Y., Liu S., Mickey S. K., Turetsky D., Gottlieb D. I., and Choi D. W. 1999. Transplanted embryonic stem cells survive, differentiate and promote recovery in injured rat spinal cord. *Nat Med* 5: 1410-1412.

24. Mimeault M., Hauke R., and Batra S. K. 2007. Stem cells: a revolution in therapeutics-recent advances in stem cell biology and their therapeutic applications in regenerative medicine and cancer therapies. *Clin Pharmacol Ther* 82: 252-264.

25. Murry C. E. and Keller G. 2008. Differentiation of embryonic stem cells to clinically relevant populations: lessons from embryonic development. *Cell* 132: 661-680.

26. Nadal-Ginard B., Anversa P., Kajstura J., and Leri A. 2005. Cardiac stem cells and myocardial regeneration. *Novartis Found Symp* 265: 142-154.

27. Nelson T. J., Martinez-Fernandez A., Yamada S., Ikeda Y., Perez-Terzic C., and Terzic A. 2010. Induced pluripotent stem cells: advances to applications. *Stem Cells Cloning* 3: 29-37.

28. Nussbaum J., Minami E., Laflamme M. A., Virag J. A., Ware C. B., Masino A., Muskheli V., Pabon L., Reinecke H., and Murry C. E. 2007. Transplantation of undifferentiated murine embryonic stem cells in the heart: teratoma formation and immune response. *FASEB J* 21: 1345-1357.

29. Pera M. F., Reubinoff B., and Trounson A. 2000. Human embryonic stem cells. *J Cell Sci* 113 (Pt 1): 5-10.

30. Perrier A. L., Tabar V., Barberi T., Rubio M. E., Bruses J., Topf N., Harrison N. L., and Studer L. 2004. Derivation of midbrain dopamine neurons from human embryonic stem cells. *Proc Natl Acad Sci U S A* 101: 12543-12548.

31. Pittenger M. F., Mackay A. M., Beck S. C., Jaiswal R. K., Douglas R., Mosca J. D., Moorman M. A., Simonetti D. W., Craig S., and Marshak D. R. 1999. Multilineage potential of adult human mesenchymal stem cells. *Science* 284: 143-147.

32. Porrello E. R., Mahmoud A. I., Simpson E., Hill J. A., Richardson J. A., Olson E. N., and Sadek H. A. 2011. Transient regenerative potential of the neonatal mouse heart. *Science* 331: 1078-1080.

33. Poss K. D., Wilson L. G., and Keating M. T. 2002. Heart regeneration in zebrafish. *Science* 298: 2188-2190.

34. Reubinoff B. E., Itsykson P., Turetsky T., Pera M. F., Reinhartz E., Itzik A., and Ben-Hur T. 2001. Neural progenitors from human embryonic stem cells. *Nat Biotechnol* 19: 1134-1140.

35. Shamblott M. J., Axelman J., Wang S., Bugg E. M., Littlefield J. W., Donovan P. J., Blumenthal P. D., Huggins G. R., and Gearhart J.D. 1998. Derivation of pluripotent stem cells from cultured human primordial germ cells. *Proc Natl Acad Sci U S A* 95: 13726-13731.

36. Singla D. K. 2009. Embryonic stem cells in cardiac repair and regeneration. *Antioxid Redox Signal* 11: 1857-1863.

37. Singla D. K. 2010. Stem cells in the infarcted heart. *J Cardiovasc Transl Res* 3: 73-78.

38.   Singla D. K, Long X., Glass C., Singla R. D., and Yan B. 2011. iPS Cells Repair and Regenerate Infarcted Myocardium. *Mol Pharm*

39.   Singla D. K. and Sun B. 2005. Transforming growth factor-beta2 enhances differentiation of cardiac myocytes from embryonic stem cells. *Biochem Biophys Res Commun* 332: 135-141.

40.   Solter D. 2006. From teratocarcinomas to embryonic stem cells and beyond: a history of embryonic stem cell research. *Nat Rev Genet* 7: 319-327.

41.   Solter D. and Knowles B. B. 1975. Immunosurgery of mouse blastocyst. *Proc Natl Acad Sci U S A* 72: 5099-5102.

42.   Soria B., Roche E., Berna G., Leon-Quinto T., Reig J. A., and Martin F. 2000. Insulin-secreting cells derived from embryonic stem cells normalize glycemia in streptozotocin-induced diabetic mice. *Diabetes* 49: 157-162.

43.   Sun N., Panetta N. J., Gupta D. M., Wilson K. D., Lee A., Jia F., Hu S., Cherry A. M., Robbins R. C., Longaker M. T., and Wu J. C. 2009. Feeder-free derivation of induced pluripotent stem cells from adult human adipose stem cells. *Proc Natl Acad Sci U S A* 106: 15720-15725.

44.   Takahashi K., Tanabe K., Ohnuki M., Narita M., Ichisaka T., Tomoda K., and Yamanaka S. 2007. Induction of pluripotent stem cells from adult human fibroblasts by defined factors. *Cell* 131: 861-872.

45.   Takahashi K. and Yamanaka S. 2006. Induction of pluripotent stem cells from mouse embryonic and adult fibroblast cultures by defined factors. *Cell* 126: 663-676.

46. Thomson J. A., Itskovitz-Eldor J., Shapiro S. S., Waknitz M. A., Swiergiel J. J., Marshall V. S., and Jones J. M. 1998. Embryonic stem cell lines derived from human blastocysts. *Science* 282: 1145-1147.

47. Thomson J. A., Kalishman J., Golos T. G., Durning M., Harris C. P., Becker R. A., and Hearn J. P. 1995. Isolation of a primate embryonic stem cell line. *Proc Natl Acad Sci U S A* 92: 7844-7848.

48. Thomson J. A. and Odorico J. S. 2000. Human embryonic stem cell and embryonic germ cell lines. *Trends Biotechnol* 18: 53-57.

49. Wagers A. J. and Weissman I. L. 2004. Plasticity of adult stem cells. *Cell* 116: 639-648.

50. Williams R. L., Hilton D. J., Pease S., Willson T. A., Stewart C. L., Gearing D. P., Wagner E. F., Metcalf D., Nicola N. A., and Gough N. M. 1988. Myeloid leukaemia inhibitory factor maintains the developmental potential of embryonic stem cells. *Nature* 336: 684-687.

51. Yu J., Vodyanik M. A., Smuga-Otto K., Antosiewicz-Bourget J., Frane J. L., Tian S., Nie J., Jonsdottir G. A., Ruotti V., Stewart R., Slukvin I. I., and Thomson J. A. 2007. Induced pluripotent stem cell lines derived from human somatic cells. *Science* 318: 1917-1920.

52. Zuk P. A., Zhu M., Ashjian P., De Ugarte D. A., Huang J. I., Mizuno H., Alfonso Z. C., Fraser J. K., Benhaim P., and Hedrick M. H. 2002. Human adipose tissue is a source of multipotent stem cells. *Mol Biol Cell* 13: 4279-4295.

Made in the USA
Lexington, KY
08 January 2013